Springer Series in Geomechanics and Geoengineering

Series editor

Wei Wu, Universität für Bodenkultur, Vienna, Austria
e-mail: wei.wu@boku.ac.at

Geomechanics deals with the application of the principle of mechanics to geomaterials including experimental, analytical and numerical investigations into the mechanical, physical, hydraulic and thermal properties of geomaterials as multiphase media. Geoengineering covers a wide range of engineering disciplines related to geomaterials from traditional to emerging areas.

The objective of the book series is to publish monographs, handbooks, workshop proceedings and textbooks. The book series is intended to cover both the state-of-the-art and the recent developments in geomechanics and geoengineering. Besides researchers, the series provides valuable references for engineering practitioners and graduate students.

More information about this series at http://www.springer.com/series/8069

Wei Wu
Editor

Recent Advances in Geotechnical Research

Springer

Editor
Wei Wu
Institute of Geotechnical Engineering
University of Natural Resources and Life Sciences
Vienna, Austria

ISSN 1866-8755 ISSN 1866-8763 (electronic)
Springer Series in Geomechanics and Geoengineering
ISBN 978-3-030-07827-0 ISBN 978-3-319-89671-7 (eBook)
https://doi.org/10.1007/978-3-319-89671-7

Printed on acid-free paper

This Springer imprint is published by the registered company Springer International Publishing AG part of Springer Nature
The registered company address is: Gewerbestrasse 11, 6330 Cham, Switzerland

Preface

This book brings together some 19 papers by members of our institute and our project partners. The book presents a snapshot of the research activities between 2014 and 2016 at our institute in Vienna. The research in this period is characterized by three research projects funded by the European Commission, i.e. MUMOLADE (Multiscale modelling of landslide and debris flow), REVENUES (Reinforced Vegetation Numerical Evaluation of Slopes) and GEORAMP (Geohazards—Risk Assessment, Mitigation and Prevention). All three projects deal with the major forms of geohazards, i.e. landslide and debris flow. Both numerical and experimental techniques are used, e.g. geotechnical centrifuge, FEM and SPH. The effect of plants on slope stabilization has received much attention. Granular mechanics and constitutive modelling remain the eternal topics in our institute.

We benefit from the staff exchange with partners in China, Europe and USA. We thank the European Commission for the financial support to the following projects:

- MUMOLADE (Multiscale modelling of landslide and debris flow), Contract Nr. 289911, Marie Curie ITN within the 7th Frame Program;
- REVENUES (Reinforced Vegetation Numerical Evaluation of Slopes), Contract Nr. 32446, Industry-Academia Partnerships and Pathways (IAPP) within the 7th Frame Program;
- GEORAMP (Geohazards—Risk Assessment, Mitigation and Prevention), Project ID: 645665, Marie Skłodowska-Curie Research and Innovation Staff Exchange (RISE) within Horizon 2020.

All members of our institute and our project partners deserve my heartfelt thanks. I wish to thank our secretary Jenny Allén for proof reading and compiling.

Vienna, Austria
October 2017

Wei Wu

Contents

Role of Plants—Results of Pore Pressure Monitoring in a Slope

M. S. Acharya

Abstract Plants on a slope can have mechanical, biological and hydrological roles which influence the stability of the slope. In most cases, soil moisture and sub-surface water play a vital role in the stability of a slope. Soil moisture, pore water pressure and the movement of surface and sub-surface water are generally dependent on the characteristics of soils and density of vegetation growing on the slope. In order to investigate the effects of plants on the soil strength, field monitoring of soil moisture and pore water pressure on a slope supported by a vegetative log crib wall was carried out in Lower Austria for a period of about 18 months. The monitoring result shows that the presence of plants stabilizes the soil moisture fluctuations during dry season, increases the soil suction and reduces the pore water pressure during the plant growth season. With the growth of plants, the rate of fluctuations of the soil moisture decreases over time. The rainfall has no significant effects on pore water pressure fluctuations during the active plant growth season.

1 Introduction

The stability of a slope depends on the strength of the soil material comprising of the slope and its geometry. Plants growing on the slope can have mechanical, biological and hydrological roles which influence the physical characteristics of the earth materials on the slope, which ultimately affect the stability. The mechanical contributions of plants mainly arise from the physical interactions of either the foliage or the root system of the plants growing on the slope [3]. The biological contribution in the soil strength may be due to soil aggregation by plant roots because a plant

M. S. Acharya (✉)
Institute of Geotechnical Engineering, University of Natural Resources and Life Sciences, Feistmantelstrasse 4, 1180 Vienna, Austria
e-mail: Madhu_sudan.acharya@boku.ac.at

W. Wu (ed.), *Recent Advances in Geotechnical Research*,
Springer Series in Geomechanics and Geoengineering,
https://doi.org/10.1007/978-3-319-89671-7_1

generally enhances the biological activities in the soil, which affects the physical and chemical characteristics of the soil and the natural drainage system of the slope, which finally has an influence on the stability. Likewise, the plants on a slope will also affect the hydrological and morphological characteristics of the slope, which has a significant influence on soil moisture balance, which again affects the soil strength.

Moreover, the plants on a slope play a vital role in the hydrological cycle and the surrounding ecosystem. Plant leaves intercept rainfall, causing absorptive and evaporative losses that reduce the amount of rainwater available for infiltration or making splash on the surface. Roots and stems increase the roughness of the ground surface and the permeability of the soil, leading to increased infiltration capacity of the ground, which generally have a positive role in slope stability. With increased permeability, it may sometimes cause depletion of soil moisture in the upper strata which may accentuate desiccation cracking in the soil resulting in higher infiltration capacity, which has a negative effect in slope stability [4]. In general, the plant roots increase the soil suction reducing pore water pressures, which again significantly increases the cohesion (c), and also the friction angle (φ) to some extent. In an experimental investigation carried out on a highway embankment in Germany, an increase in effective cohesion from 1.1 to 6.3 kN/m^2 and friction angle from 33.1° to 34.7° were observed [5]. In general, the net effect of vegetation on a slope is usually beneficial to slope stability.

In general, there are five main functions that the plants may perform within a soil bioengineering system (i.e. in a vegetative structure): support, anchor, drain, reinforce and armour. Depending upon the type of plants used in soil bioengineering, the nature of slope and soil characteristics, the plant may perform one or more of the above functions and plays a vital role in the slope stability. Reckoning the beneficial roles of vegetation on slope stability, the trend in use of vegetative systems in the form of different soil bioengineering techniques is increasing. Properly installed and maintained vegetation systems can protect and strengthen the slopes by reducing erosion, strengthening soil and inhibiting landslides.

In the above context, field monitoring of plant-soil parameters that affect the stability of the vegetated slope will be very important to understand the complex role of vegetation on slope stability. In order to investigate the effects of plants on the soil strength, field monitoring of soil moisture and pore water pressure on a slope supported by a vegetative log crib wall was carried out in Lower Austria for a period of about 18 months. The rainfall and temperature data were also collected. The results of the field measurements of soil moisture and pore water pressure within the supported backfill area of the vegetated crib wall are reported in this paper.

2 Monitoring of Slope

2.1 Objective of Slope Monitoring

As mentioned above, the soil moisture or the presence of water in soil plays an important role in the shear strength of the soil. The pore-water pressure is vital in calculating soil strengths from Terzaghi's expression for the effective stress in soil; therefore it will be one of the decisive factors that cause slope failures [2]. Soils in naturally formed slopes are generally heterogeneous. Thus, soil water processes are not only affected by precipitation and evapotranspiration, but also by texture, grain size, cracks, compactness, aggregation, roots and cavities. Therefore, the presence of plants on a slope will have a direct influence on the pore water pressure and it will keep on changing with the growth of plants and changes in ground water balance.

In this project, the monitoring of slope was done to study the long-term performance of a vegetated crib wall, which was built on a road side slope. The long-term performance of a vegetative crib wall depends on the growth of vegetation on the slope and their contribution to increase the shear strength of the soil. On the crib wall supported slope described here, rooted plants and plant's cuttings are placed into each crib layer and the pore pressure monitoring was started just after the proliferation of cuttings. Due to the growth of plants, change in hydrological characteristics of slope and other environmental factors, the soil moisture balance on a slope will change continuously, which leads to continuous changes in the pore water pressure over time. This will have direct effects on shear strength of soil and ultimately on the safety factor of the slope. As the pore water pressure changes, the safety factor of the slope will also change with time. Thus the monitoring of pore pressures, soil moisture, rainfall, temperature and growth of vegetation will provide vital information about the strength parameters of soil materials and helps to calculate safety factor of the slope more accurately.

2.2 Description of Slope

The slope monitoring site is located on the Sattelbergstrasse, a forest road in Brentenmais near Pressbaum in Lower Austria. A 5 m high vegetated wooden log crib wall was constructed on November 2003 as a retaining wall to support the road-side slope. The log crib wall is founded on 50 cm below the existing ground and has a berm in between (see Fig. 1). The wall is constructed in 10 layers of wooden log crib, each layer having an average height of about 50 cm. The total length of the wall is about 20 m. The average diameter of the wooden logs used in the construction is about 25 cm. The construction work was carried out using an excavator. Although the backfilling was done layer by layer, the backfill was not well compacted. Compaction was done just by applying the pressure on the bucket of the excavator and there was no additional equipment used for compaction.

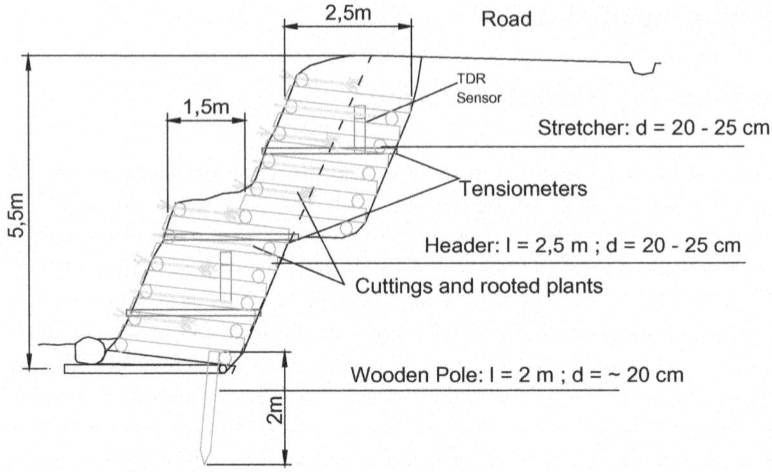

Fig. 1 Arrangements for Tensiometers and TDR sensors in the slope supported by a vegetated wooden crib wall

The backfill soil material consists of a mixture of clay, silt and gravel. Rooted plants and cuttings of five different plant species were placed horizontally into each layer of wooden crib. The length of cuttings and plants varies from 100 to 150 cm. The slope above the road is a natural forest slope (ca. 30°–35°) covered mainly with larger trees. There is a small natural ditch at the toe of the wall, which drains the water from the road and the slope above.

As the inserted cuttings and rooted plants started to proliferate, instrumentation for soil moisture, temperature and pore water pressure measurements was set up in June 2005. A continuous data collection system was installed on the slope and data were directly recorded on a data logger stationed on site [1]. The recorded data are analysed and the variations of pore water pressure and soil moisture with the growth of plants are investigated and the results are presented in this paper.

2.3 *Field Instrumentation*

Altogether nine numbers of UMS T4 Tensiometers were placed horizontally in three layers inside the vegetated log crib wall for the pore pressure measurements. The lengths of the Tensiometers used were 1.5 and 1.8 m. Shorter Tensiometers were used in the upper layers and longer ones were installed in the lower layers. In addition to these Tensiometers, four Time Domain Reflectometer (TDR) sensors were installed for soil moisture contents measurements at two different levels, and three sensors for temperature measurement were also installed at the site (Fig. 1). The details about the Tensiometer and TDR measurements are presented in the following sections.

2.4 Pore Pressure Measurement with UMS T4 Tensiometers

The Tensiometer is used to measure water tension in the soil in the non-saturated zone. In the saturated zone it acts as a piezometer and measures the water pressure in the soil. The water tension is measured and converted into a defined continuous electrical signal which is adapted to a data logging unit. Since the Tensiometer measurements are not directly influenced by the soil heterogeneity, therefore the T4 precision Tensiometer developed for outdoor monitoring projects by UMS, Germany was used for the measurement of the pore water pressure. These are reliable for outdoor use when installed deeper than 30 cm. The technical specification of UMS T4 Tensiometer used in the pore water pressure monitoring is presented in Table 1.

2.5 Moisture Content Measurement with TDR (Time Domain Reflectometers) Sensor

TDR measurement depends on discontinuities in the energy storage mechanisms of a medium. The probe tips of a TDR appliance present a discontinuity in the wave propagation path of the energy initiated at the signal source. Energy which does not become dissipated returns to its source. Combined with knowledge of the propagation velocities of the waves in the medium, these discontinuities can be located by observing the change in energy levels at fixed points in the media. Therefore, TDR sensors are widely used to measure the different physical properties of soil materials.

Four TDR sensors CS616 Manufactured by Campbell Sci., U.S.A. were used for the measurement of the soil moistures at two different levels. The CS616 Water Content Reflectometer measured the volumetric water content of porous media using time-domain measurement methods. The probe consisted of two stainless steel rods connected to a printed circuit board. A shielded 4-conductor cable was connected to the circuit board to supply power and to enable the probe and monitor the pulse

Table 1 Technical specification of UMS Tensiometer T4

Tensiometer parameters	Working ranges
Measuring range	−1000 to +850 hPa
Output signal	−100 to +85 mV
Accuracy	±5 hPa
Supply typ.	10.6 VDC/1.3 mA
Shaft Ø	25 mm; acrylic
Shaft lengths	10–200 cm
Porous ceramic	Ø 24 mm; length 60 mm, bubble point 15 bar
Cable ex corpus	1.5 m; 4 pin plug M12

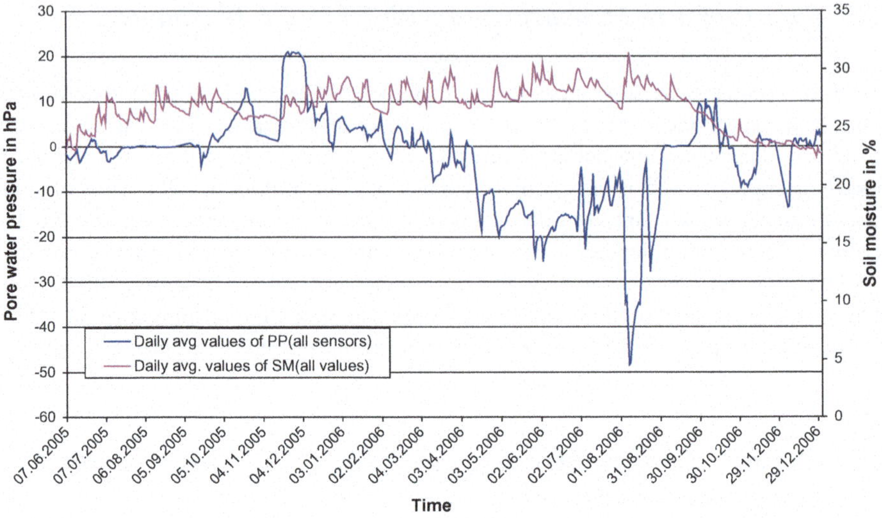

Fig. 2 The daily average values of pore water pressure and soil moistures

output. During installation, the complete probe was buried into the surface and was connected to the single-ended analogue input of the data logger.

The measurement of pore water pressure and soil moisture content were started on 7th June 2005 and continued till 31st December 2006. The data logger was set for the measurement of pore pressure, water content and temperature at an interval of 15 min.

3 Results and Discussions

The measured data were first verified and checked for errors. Any unrealistic values were deleted and then hourly and daily averages were calculated for the further analysis. The measured data values for soil moisture and pore pressure in three different levels are also compared logically for its validity. Then the validated and refined data are analysed and the results are presented in the following charts.

The chart in Fig. 2 shows the daily average values of pore water pressure and soil moisture content. The chart shows the seasonal variations in soil moisture and pore water pressure. The average pore water pressure is positive during winter and negative during summer time. This chart shows that with the growth of plants (during summer months) there is a tendency to a decrease in the absolute value and also in the fluctuation of pore water pressure, while the soil moisture remains almost unchanged.

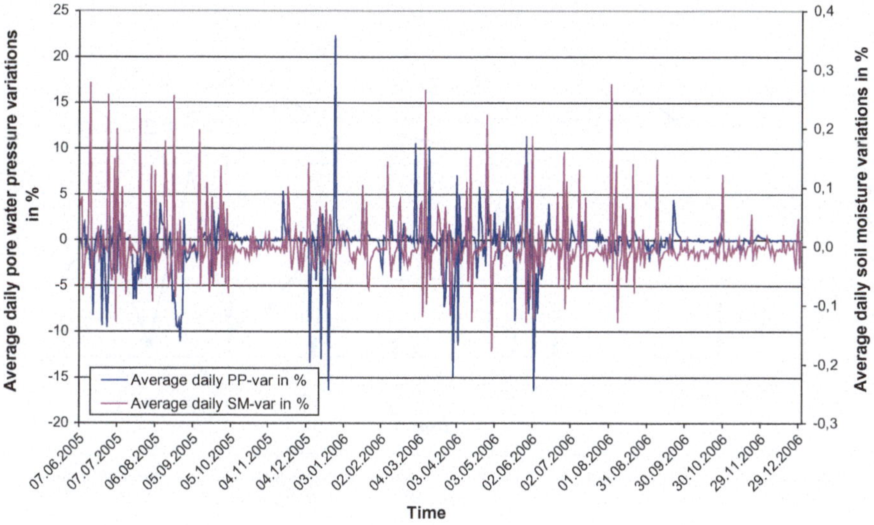

Fig. 3 The daily fluctuations of pore water pressure and soil moisture

The chart in Fig. 3 shows the daily fluctuations in pore water pressure and soil moisture content. These fluctuations are more in summer months and less in winter months which is obvious. The chart in Fig. 4 shows the monthly average variations in pore water pressure and soil moisture content. The average monthly rainfall is also presented in the chart. The comparison of pore water pressure and soil moisture variations at three different levels is presented in Fig. 5. This chart shows the daily average values of pore water pressure and soil moisture over time.

The monthly average variation of soil moisture and pore water pressure with average monthly rainfall is presented in Fig. 6. This chart shows that, although there is not a large difference in soil moisture, there are significant changes in the pore water pressure values during summer and winter months. During winter months the average monthly pore water pressure remains positive and during summer it is negative. However, the average value of soil moisture fluctuates between 23 and 29% throughout the year. There is no significant effect of rainfall on monthly average values of pore water pressure and soil moisture. The chart in Fig. 7 shows the cumulative sum of differences in average daily values of pore water pressure and soil moisture. This chart shows the net changes (fluctuations in %) in pore water pressure and soil moisture over time. The daily moisture content changes within a 10% range, while the daily difference in pore water pressure values varies from −15 to +48 hPa, i.e. up to 400%.

The chart in Fig. 7 shows that, during the plant growth season, there is a decreasing trend in pore water pressure fluctuations, and during the winter time it is in increasing trend, i.e. high fluctuations. With the growth of plants on the slope, the rate of change of pore water pressure variations (fluctuations) becomes smaller.

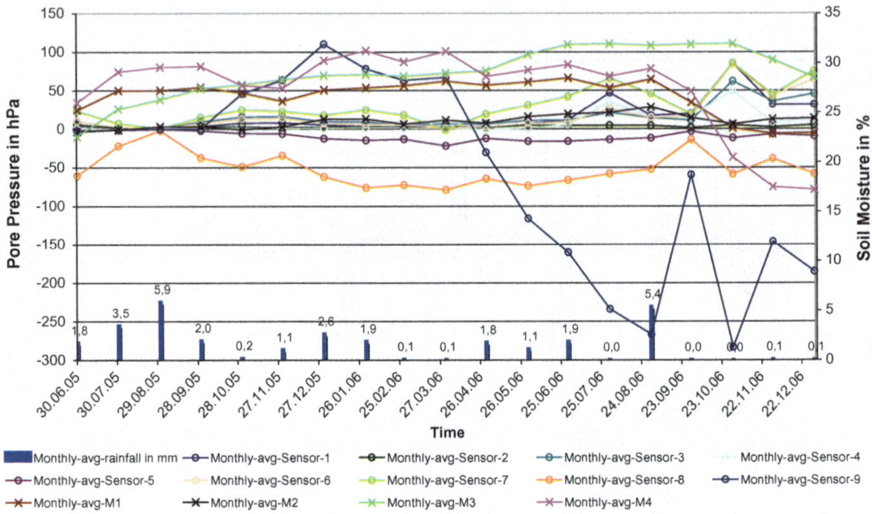

Fig. 4 The pore water pressure and soil moisture monthly average variations

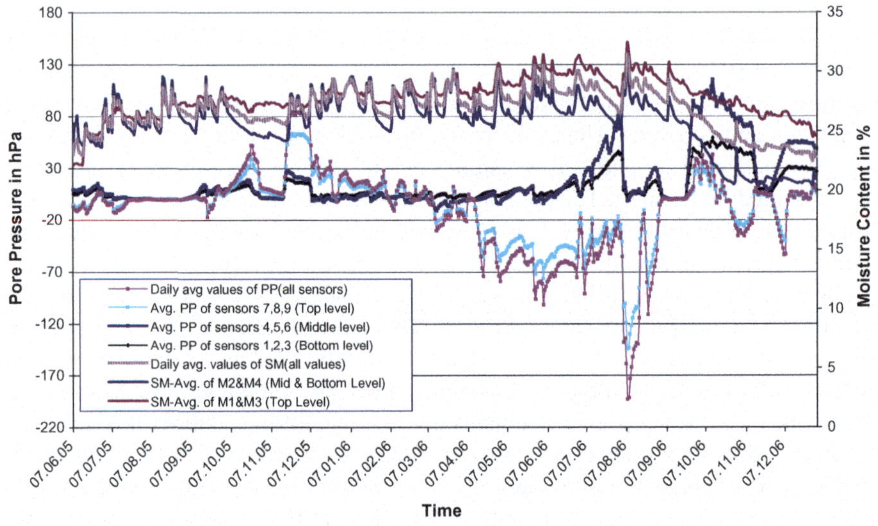

Fig. 5 The pore water pressure and soil moisture average daily variations

4 Conclusion

The field monitoring results show that the measured values of pore-water pressures are negative during summer months and they are positive during winter months. The daily fluctuations on the net values of soil-moisture and pore-water pressure are different in summer and winter months. With minor fluctuations in soil-moisture, there

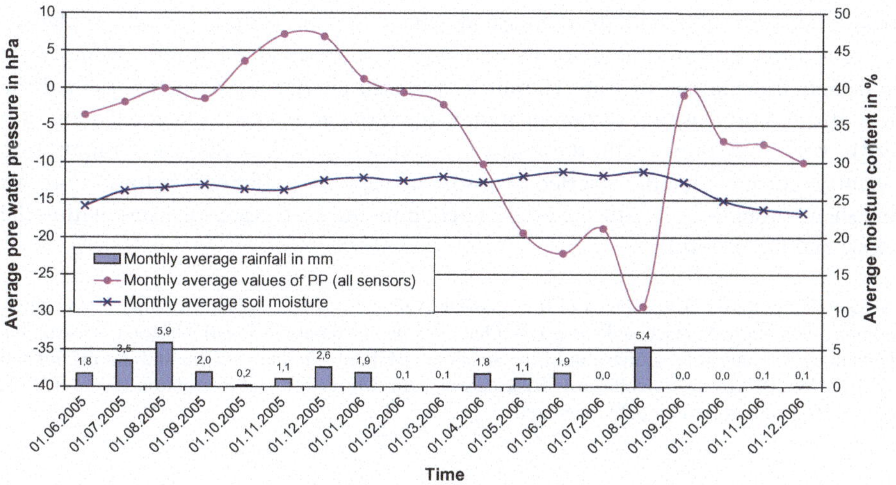

Fig. 6 The pore water pressure and soil moisture average monthly variations

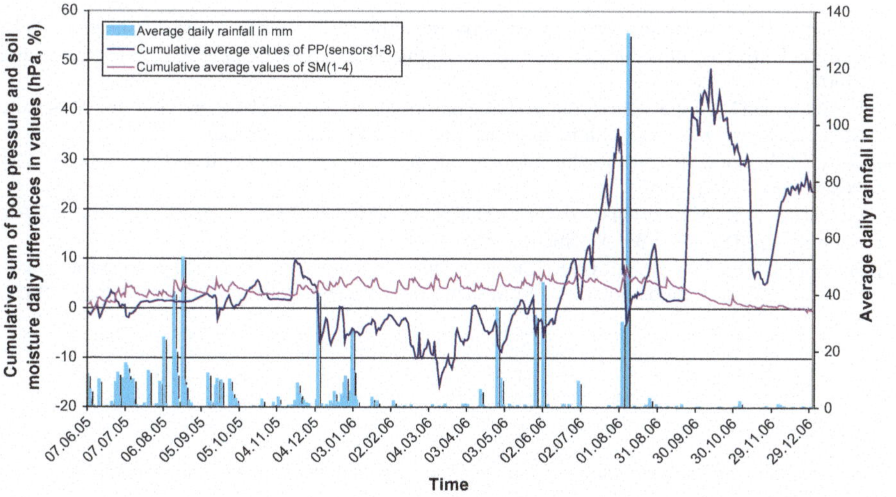

Fig. 7 The cumulative sum of difference in average daily values of pore water pressure and soil moisture variations

are very high fluctuations in pore-water pressure during summer and low fluctuations in winter. During the active growth period of the plants (spring/summer), there is a very little effect of rainfall on the pore-water pressure, whereas during the dormant season, the effect of rainfall on pore-water pressure is significantly high.

During summer months, when the most of the plant growth takes place, the pore-water pressure decreases despite of substantial rainfall events. It clearly shows that there is a positive influence of plants in reducing pore-water pressure in soil. The results also show that with the increase in plant's biomass volume (growth of plants), there is a tendency to reduce or to stabilise the pore-water pressure, soil-moisture and their fluctuations in soil. Therefore, the crib elements (wooden logs) will be less affected by seasonal variations in moisture content in case of vegetated log-crib walls, which will increase the durability of vegetated log-crib walls. These monitoring results confirmed that the inserted plants in the vegetative crib wall helps to improve the shear strength of the fill material by reducing the pore-water pressure within the slope and the backfill.

Acknowledgements The support of the Austrian Agency for International Mobility and Cooperation in Education, Science and Research (OeAD), Ebendorferstraße 7, 1010 Vienna, Austria, the University of Natural Resources and Life Sciences (BOKU), Vienna, Austria and the EC-funded HORIZON 2020 project Geohazards Risk Assessment, Mitigation and Prevention (GEO-RAMP, No. 645665) are gratefully acknowledged.

References

1. Acharya, M.S.: Suitability of vegetated wooden/bamboo crib walls for slope stabilisation. Ph.D. thesis at University of Natural Resources and Applied Life Sciences, pp. 55–72. Vienna, Austria (2007)
2. Fredlund, D.G.: Slope stability analysis incorporating the effect of soil suction. In: Anderson, M.G., Richards, K.S. (eds.) Slope Stability, pp. 113–143. Wiley, Chichester (1987)
3. Gray, D.H., Sotir, R.B.: Biotechnical and Soil Bioengineering Slope Stabilisation, pp. 1–129. Wiley, New York (1996)
4. Greenway, D.R.: Vegetations and slope stability. In: Anderson, M.G., Richards, K.S. (eds.) Slope Stability, pp. 187–230. Wiley, Chichester (1987)
5. Katzenbach, R., Werner, A.: Experimentelle Untersuchungen zur Standsicherheitsproblematik von historischen Bahndämmen, 5. Österreichische Geotechniktagung, Vienna, Austria (2005)

A Probabilistic 3-D Slope Stability Analysis for Forest Management

Alessio Cislaghi, Chiara Vergani, Enrico Antonio Chiaradia and Gian Battista Bischetti

Abstract A 3-D physical-based approach to slope stability has been proven to be very promising in order to provide reliable spatially distributed landslides maps. Over large areas, however, such an approach still presents some limitations, mainly related to the variability and the uncertainty of the input parameters. By combining a 3-D physical-based model with a Monte Carlo technique, such constraints can be overcome, improving the performance and the applicability of the method. Whereas uncertainties of geotechnical, morphological and hydrological parameters have been widely investigated, few studies have been focused on the variability of root reinforcement, which plays a crucial role in preventing shallow landslides in forested areas. To contribute to define the effect of different forest management strategies on slope stability, we developed a 3-D model able to properly take into account for the effect of the root systems into the soil. The main objectives of our study are: (i) to define a probability distribution function for the root reinforcement according to the forest stands characteristics (tree density, mean diameter at breast height, minimum distance between trees), (ii) to obtain a probability distribution of the Factor of Safety through the combination between a 3-D slope stability model and a Monte Carlo simulation technique, and (iii) to evaluate the difference between several for-

A. Cislaghi (✉) · E. A. Chiaradia · G. B. Bischetti
Department of Agricultural and Environmental Sciences,
Università degli Studi di Milano, via Celoria 2, 20133 Milan, Italy
e-mail: alessio.cislaghi@unimi.it

E. A. Chiaradia
e-mail: enrico.chiaradia@unimi.it

G. B. Bischetti
e-mail: bischetti@unimi.it

C. Vergani
Department of Forest Science, Bern University of Applied Sciences,
Langgasse 85, 3052 Zollikofen, Switzerland
e-mail: chiara.vergani@bfh.ch

W. Wu (ed.), *Recent Advances in Geotechnical Research*,
Springer Series in Geomechanics and Geoengineering,
https://doi.org/10.1007/978-3-319-89671-7_2

est management scenarios in terms of shallow landslide risk. The model has been applied to a small Alpine area, mainly covered by coniferous forest and characterized by steep slopes and a high landslide hazard. Our findings contribute to provide forest managers with useful information for understanding the consequences of different forestry strategies.

1 Introduction

Forests have a stabilizing effect on slope stability, especially in mountainous terrains [1]. Indeed, vegetation influences both hydrological processes, which affect the water content into the soil and consequently the pore pressure, and the mechanical mechanisms of soil reinforcement. In recent decades, several scientific works studied and quantified the beneficial effect of the root systems on the soil strength, commonly defined root reinforcement. It is now well-known that the intensity of such reinforcement mainly depends on root distribution and on root tensile strength [2–4], which are related to forest stand characteristics [5–7]. However, few studies have been focused on the relationships between slope stability, root reinforcement and forest stand characteristics, such as the trees density, the diameter at breast height of the trunk DBH, the trees age, etc.

Root reinforcement should be an input parameter of the physical-based slope stability models used to analyse and predict slope instabilities in forested landscapes. Most of these models are based on the infinite slope theory, a simplified and well-known low dimensional approach that assumes a plane surface of failure parallel to the ground and ignores the stresses acting on the four lateral sides of the sliding block because they balance each other [8]. Actually, such approach has been used to reliably and suitably define shallow landslide occurrences when the topography has a major control on the shallow movements [9], but many authors emphasized some limitations [10–12] that tend to overestimate the slope instabilities [13]. In addition, the infinite slope assumptions do not allow to account for the role of root reinforcement on lateral sides of the shearing block. Finally, physical-based models request a deep knowledge of all input parameters whose measure and/or evaluation incorporate a certain degree of uncertainties and spatial variability. To overcome these disadvantages, a multidimensional slope stability approach combined with the Monte Carlo Simulation (MCS) that adopts different probability distribution function for each input parameters, has been developed providing reasonable results [14].

While the probability distribution functions of the geotechnical properties have been subject to several analyses [11, 15, 16], the spatial and the temporal variability of the root reinforcement is still an open issue [11, 15]. Moreover, such variability can be associated with the forest dynamics, which are strongly influenced by the forest management practices. Indeed, sustainable forest management requires to establish decisions on the regeneration method, species composition, forest structure (e.g. growth, development, and spacing) through the thinning interventions [17]. Forestry operations, in fact, generally aim to maximise the economic returns, to provide better conditions in order to produce quality timber. At the same time, however, they should guarantee a canopy cover able to prevent soil erosion and landslides on steep slopes.

Against this background, this study applies a modular approach, based on a 3-D stability model coupled with MCS and root reinforcement models for estimating slope stability after different forestry interventions. Although the types and the intensity of such operations are numerous, in this study we considered only the most common thinning and cutting strategies and applied the method to a small subalpine catchment covered by a coniferous forest.

2 Materials and Methods

2.1 Probabilistic 3-D Slope Stability Model

A 3-D approach to hillslope stability modelling was selected to overcome the drawbacks of infinite slope analysis. The model is the multidimensional shallow landslide model, MD-STAB, developed by Milledge et al. [14]. Based on the pioneering work of Burroughs [18], it is a simplified 3-D limit equilibrium force balance on a slope block considering all the forces acting on each face of the block. The main assumptions of MD-STAB are as follows: (i) the force balance is applied to the central block, (ii) the groundwater level is steady and parallel to the slope surface, (iii) the model ignores infiltration, suction and capillary rise effects in the unsaturated zone, and (iv) the single block is partitioned into saturated and unsaturated regions. For each block, the model provides the Factor of Safety FS that can be evaluated as the ratio between the resisting and driving forces:

$$FS = \frac{F_{rb} + 2\ F_{rl} + F_{rd} - F_{du}}{F_{dc}} \tag{1}$$

where the driving forces consist of the downslope component of the central block mass, F_{dc}, and the force acting on the central block from the upslope wedge, F_{du}, assuming active earth condition. The resisting forces acting on all boundaries of the block are due to the passive earth pressure from the downslope wedge, F_{rd}, the basal resistance force, F_{rb}, and the shear resistance on the two parallel slope sides of the element, F_{rl}. More details are provided in the work of Milledge et al. [14]. The geometrical and morphological parameters (width, length and slope of the analysed block) were evaluated from a digital elevation model and from a landslides inventory, whereas the geotechnical properties and the root reinforcement were estimated from the probability distribution functions through a MCS repeated at least 1000 times as suggested by Hammond et al. [15]. Such a procedure provides a probability distribution of FS values for each block and, consequently, calculates the probability of failure, which is the probability that FS is less than the unity.

The performance of the proposed probabilistic 3-D slope stability model have been evaluated through the Receiver-Operating Characteristic (ROC) curves [19] and, in particular, the Area Under the ROC Curve AUC [6].

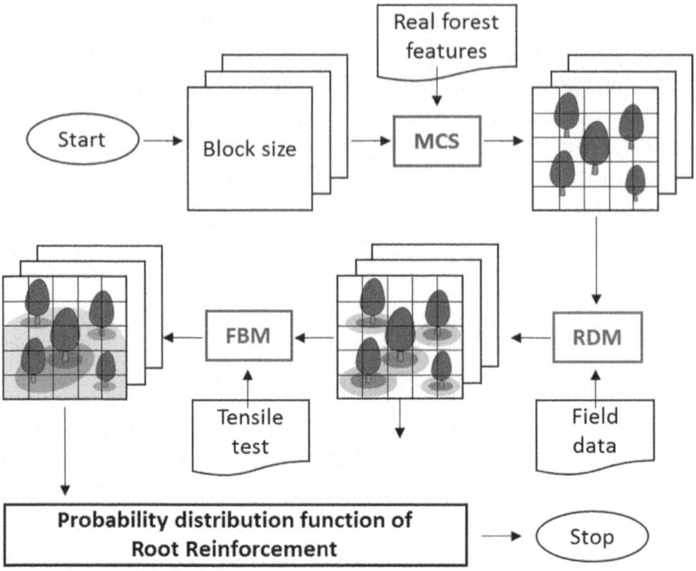

Fig. 1 Flowchart for the evaluation of the probability distribution function of root reinforcement

2.2 Probability Distribution Function of the Root Reinforcement

Root reinforcement is highly dependent on the forest stand characteristics, such as the spatial distribution of root systems [6, 20, 21] and is a function of the root density of different diameter classes and of the ultimate resisting force before the rupture [1, 4]. On these premises, we developed a specific multi-steps method summarized in the flow chart (Fig. 1). Starting from real stand features (e.g., density of trees, average DBH, minimal distance between trees), a set of random forest configurations can be achieved by MCS in each cell. The procedure generates random locations of trees (couples of coordinates) according to the minimal distance and tree density. For each forest configuration, the Root Distribution Model RDM developed by Schwarz et al. [22] was applied to evaluate the spatial distribution of root diameter for each cell as input. A calibration of the main parameters is necessary, minimizing the discrepancies between field measurements (performed by digging three profiles at different distances from the stem) and the simulated root densities. Furthermore, the root reinforcement model combines the root density of different diameters and the force-diameter relationship obtained from laboratory or field tensile tests, to evaluate the root reinforcement. In this case, we chose the Fiber Bundle Model (FBM) proposed and described by Pollen and Simon [23]. The result of FBM implementation is a root reinforcement map and an empirical probability distribution function.

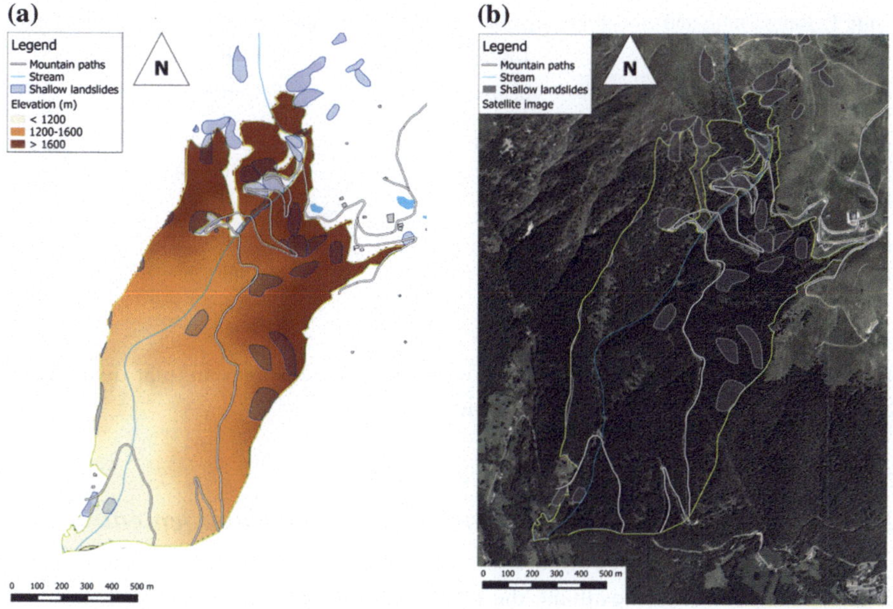

Fig. 2 **a** Digital elevation model, and **b** satellite image and landslide inventory of the Lower East D'Ornica catchment

2.3 Case Study and Model Application

The proposed method was tested in the Lower East D'Ornica catchment, a subalpine tributary of the Brembo River (Central Alps, Lombardy, North Italy). The study area is 1.1 km^2, the steepness reaches 48° and the elevation ranges from 1100 to 1730 m a.s.l. (Fig. 2a). The area is completely covered by a mixed forest of European silver fir (*Abies alba* Mill.) and Norway spruce (*Picea abies* (L.) Karst), which are the predominant species (60–70 and 20–30% respectively) and is located in the Mesalpic Forest Region. The forest stand is at least 80 years old. The basal area of the living stand was 30.36 m^2/ha and the density of trees is approximately 400 trees/ha. The mean height and the diameter at breast height are 31 and 0.31 m, respectively. The area has a rainy weather, with mean annual precipitation equal to 1900 mm, mainly concentrated in autumn. The soils are classified as Cambisols-Podzols on sandstone, siltstone and mudstone according to the WRB soil classification. The texture can be classified as clean sand of silt grains to a depth of 150 cm. The landslides cover approximately 8% of the area and the inventory reported that shallow mass movements such as rotational and transitional shallow landslides and debris flows as shown in Fig. 2b.

The digital elevation model is available with a resolution of 10 m × 10 m. The input parameters of the proposed model were set as uniform distribution for the geotech-

Table 1 Range, units and probability distribution functions of input parameters used in MD-STAB

Variable	Min-Max (Apex)	Units	PDF
ϕ	24–28	degrees	Uniform
γ_s	16–18	KN/m^3	Uniform
Z	0.8–1.1 (1)	m	Triangular
W	2–80	m	Weibull
L	10–300	m	Weibull
R/T	10^{-4}–10^{-1}	m^{-1}	Uniform

nical properties and the hydrological conditions, whereas as triangular distribution for the soil depth as summarized in Table 1.

2.4 *Evaluating Different Scenarios of Forest Management*

The procedure allows to evaluate the effects of different forestry interventions on the root reinforcement, and then on the slope stability, for a period of ten years after cutting condition, assuming a complete decay of the root systems [24, 25].

Thinning is the main method for influencing the growth and the development of the trees. According to Kerr and Haufe [17], any thinning or cutting can be described in terms of thinning type, thinning intensity and distribution of remaining trees. There are two main types of thinning: the low thinning or "thinning from below" and the crown thinning or "thinning from above". The first aims to remove the lower canopy (i.e. suppressed, sub-dominant trees, trees with smaller DBH) and to concentrate potential for growth on the larger diameter trees, whereas the other aims to remove trees from the dominant crown classes in order to favour the best trees of the same crown size. In addition, a common approach is the intermediate thinning that combines the other two. The thinning intensity is strongly influenced by the characteristic of re-growth of the forest and it is difficult to quantify for a generic coniferous forest. Concerning the distribution of trees, uniform stands are clearly the best solution from a timber perspective, but not in terms of habitat diversity and resilience after disturbances. Basing on this description, we set three scenarios (Table 2) and, in addition, we run a simulation considering a clear-cutting approach (A000) in order to quantify the beneficial role of the forests on slope stability and to identify which areas are less prone to the instabilities. Such areas are suitable for the gap-cutting approach that is common in monocultures of conifers, improving the natural regeneration, diversifying the tree species and guaranteeing economic sustainability for forest companies.

3 Results

3.1 Root Reinforcement

For the proposed procedure, we needed to calibrate the incorporated models, RDM and FBM. In nine different trenches, roots were counted at three different distances from the tree stems (0.5, 1.5 and 2.5 m) and were classified according to the diameter classes [3]. The four parameters of RDM were calibrated minimizing the differences between the simulated and observed root density at the experimental site. Tensile tests in laboratory were carried out, providing the relationship between the force (F) and the diameter (ϕ) [1, 2] in order to apply the FBM.

$$F = 4.78\ \phi^{2.04} \tag{2}$$

Once RDM and FBM have been calibrated, the MCS generated 1000 virtual forest stand configurations for each cell, making available 1000 different probability distributions of root reinforcement. The base scenario A400 provided an average root reinforcement of 7.35 kPa. On average, 10% of the values are lower than 5 kPa, whereas around 45% of the root reinforcement ranges from 5 to 10 kPa.

Considering the post-intervention scenarios, the values of root reinforcement for the same thinning type with different intensity produced rather similar results, whereas greater differences were obtained as function of the type of thinning (Fig. 3).

Crown thinning seems to cause an intense decrease of root reinforcement for both thinning intensities: the median values of root reinforcement are 2.39 and 1.57 kPa for C320 and C260 respectively. On the contrary, low thinning generates an increase of root reinforcement: the median values increase up to 10.78 and 7.93 kPa for L320 and for L260. Such results can be ascribed to a spreading of the root systems of the

Table 2 Different scenarios combining thinning type and thinning intensity after 10 years from the cutting

Thinning type	Thinning intensity	
	Light thinning (removing 20% of the trees)	Moderate thinning (removing 35% of the trees)
Low thinning (increasing of 20% the average DBH)	L320	L260
Intermediate thinning (maintaining the average DBH)	I320	I260
Crown thinning (decreasing of 20% the average DBH)	C320	C260

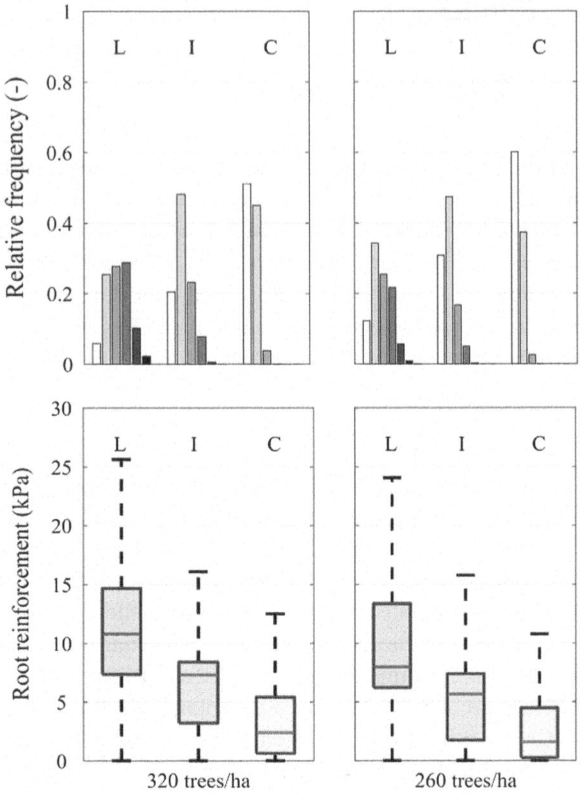

Fig. 3 Root reinforcement values for different scenarios

surrounding, more robust trees, which colonize the soil left free by the felled trees, which root systems decay.

In the case of intermediate thinning, which maintains the same distribution of DBH, there is a significant increase of the probability that the root reinforcement values are lower than 5 kPa (20% for I320 and 30% for I260) with respect to the condition before the cutting. The median values of root reinforcement are 7.28 and 5.63 kPa for I320 and I260 simply due to the reduction of the tree density.

3.2 Slope Stability

The method was applied to provide a landslide hazard map and to evaluate the probability of failure Pr[FS < 1] for the different scenarios. The areas with a high probability of failure are located in the Eastern side of the catchment and, in particular, along the hairpins in the mountain roads (Fig. 4a). The results are in agreement with the shallow landslides recorded in the inventory. The good performance of the model

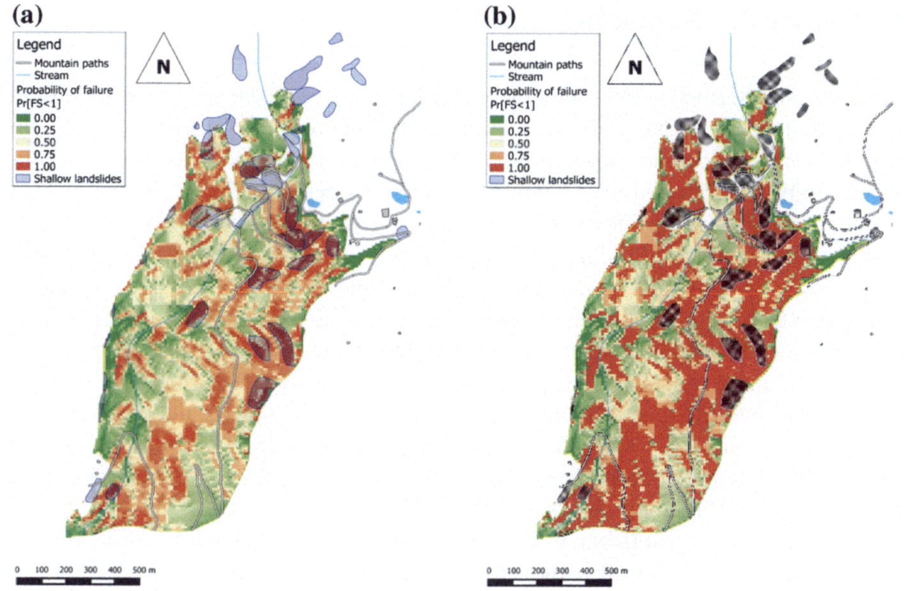

Fig. 4 MD-STAB probabilistic landslide hazard map for the base scenario A400 (**a**), and for the clear-cutting scenario A000 (**b**)

was confirmed by the balance between the successful and true negative rates, and by a good AUC value (0.793).

In general, approximately 55% of the study area has a probability of failure greater than 0.5, and 25% has a probability of failure greater than 0.75. This is mainly due to the great steepness of the hillslopes and the results of the clear-cut scenario (A000) showed a significant increase of the probability of failure, highlighting the beneficial role of root reinforcement (Fig. 4b).

The influence of the thinning types is significant in terms of the average value of root reinforcement and consequently on the slope stability in terms of probability of slope failure with respect to the present situation.

Crown thinning, however, resulted to be less adequate for the steep slopes in the Eastern part of the study area, where low thinning or single tree's selection should be suggested.

The thinning intensity also had a slight effect on the probability of failure and did not significantly modify the landslides risk of the catchment.

Although the root reinforcement is generally important for the entire area, the stability of the Western part bordered by the mountain path appeared less affected by such reinforcement, and thus suitable for gap-cutting approaches and in general for clear-cutting.

4 Conclusive Remarks

The method described in this study incorporates the more recent advances in the field of soil reinforcement due to the vegetation, and combines a 3-D stability model with a probabilistic procedure through the MCS. The method has been applied in a sub-alpine study area covered by a coniferous forest. Its performance is comparable, or even better than other physically based models and multivariate analyses. The robustness and the accuracy of the method suggest its application for a sustainable forest management in order to evaluate the effects on slope stability of different scenarios of forest interventions in terms of typology and intensity. A quantitative analysis of the consequence of forest management strategies, in fact, is a necessary step to carry out good thinning practice. The application of the method to the study case showed that low and intermediate thinning, independently of the thinning intensity, cause a small decrease of root reinforcement, and then a negligible increase of probability of slope failure. On the contrary, crown thinning causes a significant reduction of root reinforcement and then of slope stability.

These results suggest to carry out specific analyses to define the location and the intensity of the forest cutting, especially where slopes are steep.

In conclusion, a sustainable forest management, which must not lead to an abrupt reduction of slope stability, should be supported by a robust procedure in order to evaluate the suitable location, type and intensity of the thinning strategies. The presented method is an example of such a procedure.

References

1. Bischetti, G.B., Chiaradia, E.A., Epis, T., Morlotti, E.: Root cohesion of forest species in the Italian Alps. Plant Soil **324**, 71–89 (2009). https://doi.org/10.1007/s11104-009-9941-0
2. Bischetti, G.B., Chiaradia, E.A., Simonato, T., Speziali, B., Vitali, B., Vullo, P., Zocco, A.: Root strength and root area ratio of forest species in Lombardy (Northern Italy). Plant Soil **278**, 11–22 (2005). https://doi.org/10.1007/s11104-005-0605-4
3. Genet, M., Kokutse, N., Stokes, A., Fourcaud, T., Cai, X., Ji, J., Mickovski, S.: Root reinforcement in plantations of *Cryptomeria japonica* D. Don: effect of tree age and stand structure on slope stability. For. Ecol. Manag. **256**, 1517–1526 (2008). https://doi.org/10.1016/j.foreco.2008.05.050
4. Schwarz, M., Giadrossich, F., Cohen, D.: Modeling root reinforcement using a root-failure Weibull survival function. Hydrol. Earth Syst. Sci. **17**, 4367–4377 (2013). https://doi.org/10.5194/hess-17-4367-2013
5. Mao, Z., Bourrier, F., Stokes, A., Fourcaud, T.: Three-dimensional modelling of slope stability in heterogeneous montane forest ecosystems. Ecol. Model. **273**, 11–22 (2014). https://doi.org/10.1016/j.ecolmodel.2013.10.017
6. Moos, C., Bebi, P., Graf, F., Mattli, J., Rickli, C., Schwarz, M.: How does forest structure affect root reinforcement and susceptibility to shallow landslides?: a case study in St. Antönien, Switzerland. Earth Surf. Process. Landf. **41**, 951–960 (2016). https://doi.org/10.1002/esp.3887
7. Roering, J.J., Schmidt, K.M., Stock, J.D., Dietrich, W.E., Montgomery, D.R.: Shallow landsliding, root reinforcement, and the spatial distribution of trees in the Oregon Coast Range. Can. Geotech. J. **40**, 237–253 (2003). https://doi.org/10.1139/t02-113

8. Taylor, D.W.: Fundamentals of soil mechanics. Soil Sci. **66**, 161 (1948)
9. Dietrich, W.E., Bellugi, D., Real De Asua, R.: Validation of the shallow landslide model, SHALSTAB, for forest management. Land use watersheds Hum. Influ. Hydrol. Geomorphol. Urban For. Areas. pp. 195–227 (2001)
10. Casadei, M., Dietrich, W.E., Miller, N.L.: Controls on shallow landslide size. Debris-Flow Hazards Mitig. Mech. Predict. Assess. pp 91–101 (2003)
11. Haneberg, W.C.: A rational probabilistic method for spatially distributed landslide hazard assessment. Environ. Eng. Geosci. **10**, 27–43 (2004)
12. Wu, W., Sidle, R.C.: A distributed slope stability model for steep forested basins. Water Resour. Res. **31**, 2097–2110 (1995)
13. Yu, H.S., Salgado, R., Sloan, S.W., Kim, J.M.: Limit analysis versus limit equilibrium for slope stability. J. Geotech. Geoenviron. Eng. **124**, 1–11 (1998)
14. Milledge, D.G., Bellugi, D., McKean, J.A., Densmore, A.L., Dietrich, W.E.: A multidimensional stability model for predicting shallow landslide size and shape across landscapes: predicting landslide size and shape. J. Geophys. Res. Earth Surf. **119**, 2481–2504 (2014). https://doi.org/10.1002/2014JF003135
15. Hammond, C., Hall, D., Miller, S., Swetik, P.: Level I Stability Analysis (LISA) Documentation for Version 2.0 (General Technical Report No. INT-285). USDA Forest Service Intermountain Research Station (1992)
16. Zhou, G., Esaki, T., Mitani, Y., Xie, M., Mori, J.: Spatial probabilistic modeling of slope failure using an integrated GIS Monte Carlo simulation approach. Eng. Geol. **68**, 373–386 (2003)
17. Kerr, G., Haufe, J.: Thinning practice: a silvicultural guide. For. Comm. 1 (2011)
18. Burroughs, E.R.: Landslide hazard rating for portions of the Oregon Coast Range. In: Proceedings of the Symposium on Effects of Forest Land Use on Erosion and Slope Stability (May 7–11, 1984). University of Hawaii, Honolulu, pp. 265–274 (1985)
19. Frattini, P., Crosta, G., Carrara, A.: Techniques for evaluating the performance of landslide susceptibility models. Eng. Geol. **111**, 62–72 (2010). https://doi.org/10.1016/j.enggeo.2009.12.004
20. Burroughs, E.R., Thomas, B.R.: Declining root strength in Douglas-fir after felling as a factor in slope stability. USDA For. Serv. Intermt. For. Range Exp. Stn. Research Paper INT-190 (1977)
21. Neuhäuser, B., Damm, B., Terhorst, B.: GIS-based assessment of landslide susceptibility on the base of the weights-of-evidence model. Landslides **9**, 511–528 (2012). https://doi.org/10.1007/s10346-011-0305-5
22. Schwarz, M., Lehmann, P., Or, D.: Quantifying lateral root reinforcement in steep slopes - from a bundle of roots to tree stands. Earth Surf. Process. Landf. **35**, 354–367 (2010). https://doi.org/10.1002/esp.1927
23. Pollen, N., Simon, A.: Estimating the mechanical effects of riparian vegetation on stream bank stability using a fiber bundle model. Water Resour. Res. **41**, W07025 (2005). https://doi.org/10.1029/2004WR003801
24. Sidle, R.C., Bogaard, T.A.: Dynamic earth system and ecological controls of rainfall-initiated landslides. Earth-Sci. Rev. **159**, 275–291 (2016). https://doi.org/10.1016/j.earscirev.2016.05.013
25. Vergani, C., Schwarz, M., Soldati, M., Corda, A., Giadrossich, F., Chiaradia, E.A., Morando, P., Bassanelli, C.: Root reinforcement dynamics in subalpine spruce forests following timber harvest: a case study in Canton Schwyz, Switzerland. CATENA **143**, 275–288 (2016). https://doi.org/10.1016/j.catena.2016.03.038

Effect of Eccentricity and Boundary on Jet Formation of Soft Fine Sand

Deshan Cui, Wei Wu, Wei Xiang, Qiong Chen and Shun Wang

Abstract Very soft fine quartz sand is prepared in a fully decompactified state by using a sieve. After flattening the sand surface, a steel ball is dropped on the sand bed. On impact of the steel ball, fine soft sand is blown away in all directions and a transient impact crater forms. In the process of splashing, the soft sand surface first rises and then falls under the function of surface wave and air pressure. When the sand surface falls, a granular jet emerges and is driven straight into the air. After reaching the highest point, the jet goes downwards into the soft sand, thus leaving a central sand pile on the ground. Then, after a while, the air bubble beneath the sand horizontal level rises slowly towards the surface, causing a granular explosion. In addition, the eccentricity of falling point, lateral boundary material and gas permeability of bottom boundary are changed to investigate the influence on the formation of the jet.

1 Introduction

Only 128 impact craters exposed at Earth's surface have been found so far, while new impact craters are discovered occasionally [1]. A lot of information about the impact craters has been extracted from remote observations on planetary surface [2–4]. However, the impact crater scale is enormous and the direct observations are impossible. As a complement to the theoretical approach, the small scale experimental work and numerical simulation have become an important component to investigate asteroid collisions and impact processes. Thoroddsen and Shen [5] performed small scale

D. Cui (✉) · W. Xiang · Q. Chen
Faculty of Engineering, China University of Geosciences, Wuhan, China
e-mail: cuideshan@cug.edu.cn

W. Wu · S. Wang
Institute of Geotechnical Engineering, University of Natural Resources and Life Sciences, Vienna, Austria
e-mail: wei.wu@boku.ac.at

W. Wu (ed.), *Recent Advances in Geotechnical Research*,
Springer Series in Geomechanics and Geoengineering,
https://doi.org/10.1007/978-3-319-89671-7_3

experiments by letting a solid lead sphere fall onto a deep flat layer of granular medium consisting of spherical glass beads. They found a generation of a granular jet from the impact site. Lohse et al. [6] performed numerical simulations to find out the process of jet form.

As a matter of fact, the splash and jet phenomena are long known when a ball or a fluid droplet impacts on a liquid surface [7–10]. In a liquid, the splash event originates from the formation of a cavity in the process of impact and is determined by the wettability of the impacting body. Besides the deep closure of the cavity due in particular to hydrostatic pressure, the corona shaped splash above the sand surface is pulled towards the symmetry axis. Jets are found to occur regularly and have great relationship with deep and surface closures of the cavity [11].

The impact craters and jet formations of a steel sphere into a bed of soft fine sand are real complex motions. Here, we studied the role of eccentricity of falling point, lateral boundary material and air permeability of bottom plate in shaping a granular jet, and the distribution of air pressure to learn more about the subtleties of the interaction between sand and air.

2 Experimental Programme

The experimental setup (Fig. 1) consists of a transparent acrylic container of 20 cm high and 20 cm in diameter filled with extremely soft fine sand (average grain size of about 90 μm; grains are quartz sand and non-spherical). It is hard to achieve quantitatively reproducible results, due to the random force-chain networks in the quick sand [6]. Therefore, in order to prepare a well-defined initial state and decrease the influence on the lateral boundary, the fine sand was decompactified and homogenized by freely falling from a 0.4-mm mesh sieve and then flattened. The grains could freely fall into the cylinder under gravity (loose density $\rho = 1.22$ g/cm^3 and initial void ratio $e_0 = 0.99$) and settle in an extremely loose packing with the force chains substantially weakened. Under these packing, the granular is unable to support any weight. Lohse et al. [12] called this state of sand "dry quick sand".

A vacuum pump connected to the top funnel is used to keep and release the steel sphere from an initial height H, with zero initial velocity and spin. The dynamics of the sand are observed with a digital high-speed camera (500 frames per second with 200×300 pixel). Finally, top and lateral pictures of the resulting impact crater were taken to determine the dimensions.

The projectile is a steel sphere, 2.54 cm in diameters, dropped from constant height onto the sand surface, corresponding to an initial velocity of 6.64 m/s.

3 Experimental Results

3.1 Global Observations

When the impact experiment is conducted at ambient pressure, typical stages of visible events are as follows (Fig. 2): First, the steel ball falls and vanishes in the soft sand and a corona splash is created. Inhomogeneities develop in the corona, due to the inelastic particle-particle interaction and non-uniform particle size distribution. Then, the sand surface rises to the maximum height. As the sand level falls, a collimated jet shoots out from the sand at the position of impact. While the upper part of the jet goes upwards, in the lower parts the inelastic particle-particle collisions lead to density inhomogeneities in the jet. Then, the jet collapses under gravity. Finally, sand eruptions are seen at the position of impact, resembling boiling water.

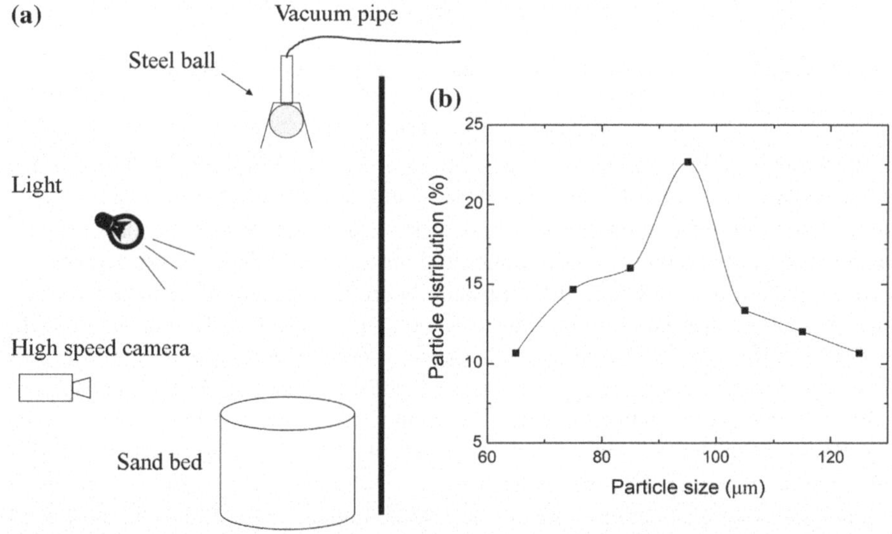

Fig. 1 **a** Schematic view of the experimental setup. **b** Particle size distribution of fine sand

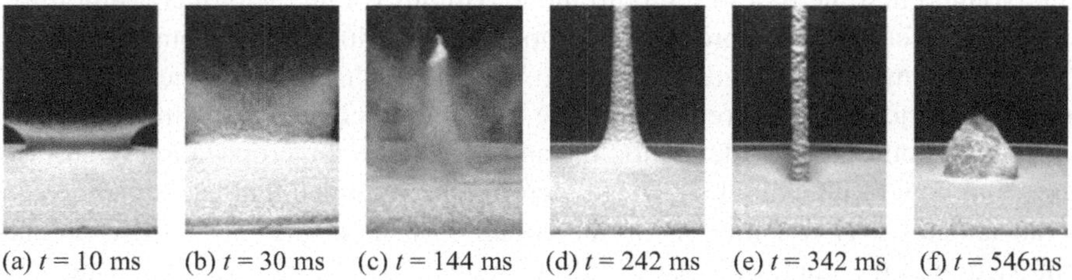

Fig. 2 Typical stages of the impact crater and jet formation: **a** corona splash, **b** rising sand surface, **c** collimated jet emergence, **d** maximum jet height, **e** collapsing jet, **f** sand eruption

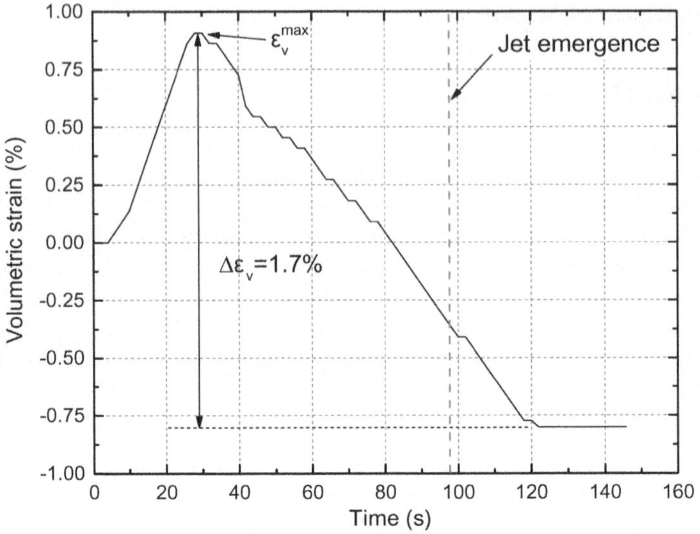

Fig. 3 The volumetric strain versus time after the impact of the steel ball on loose fine sand

But how does the volumetric change produce the jet formation? We believe that the propagation of surface wave and flow of interstitial air which the ball creates while penetrating the sand further fluidize the sand bed. When the ball moving through the sand bed there is more interstitial air to levitate the sand surface and the wave propagation would be more important. With the function of surface wave and interstitial air, a surprisingly vigorous jet shoots up from the surface of the sand. In order to test this idea we studied the change of sand bed surface after the impact of the ball by imaging a small region at the center of the container. As the sand surface is found to rise evenly in the container, the images measured with the camera give us an estimate of the sand surface height h_t at time t. $\Delta h = h_t - h_{t=0}$, where $t = 0$ is the moment of impact. The volumetric strain ε_v is defined as $\varepsilon_v = \Delta h A / V_0$, in which A and V_0 denote the sectional area of acrylic container and initial volume of sand bed, respectively.

Figure 3, in which volumetric strain ε_v is plotted versus time, shows that the sand volume first increases rapidly to the maximum volume with increasing time, and then decreases. In the process of compaction of volume, the jet emerges. After that, the volume of sand bed continues to compact. When the volumetric strain is stable, the sand jet still increases until the maximum height. That is in accordance with our hypothesis of the levitation of sand bed due to the propagation of surface wave and flow of interstitial air. It can therefore be excluded that jet formation can be attributed solely to variations in the collapse of the cavity. The expansion and compaction of the sand bed must therefore be related to the jet formation. One then may also wonder about the influence of the drop point, the lateral boundary material and the gas permeability of bottom boundary on the jet formation.

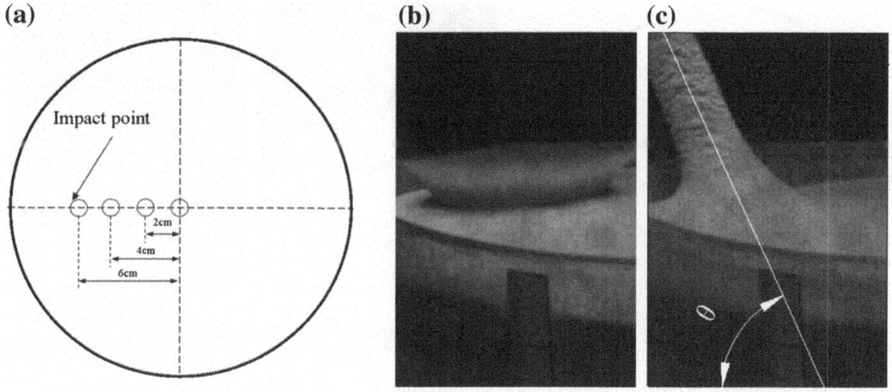

Fig. 4 **a** Schematic view of the eccentricity test. **b–c** Splash and inclined jet at 6 cm eccentricity

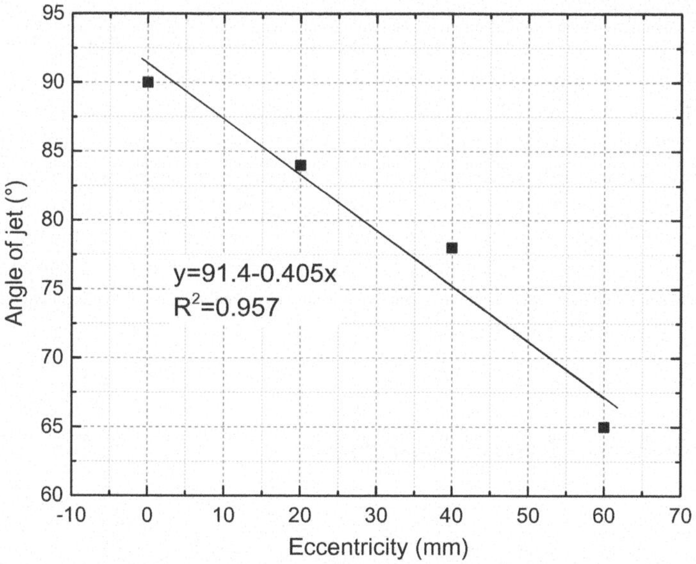

Fig. 5 Angle of jet as a function of eccentricity for steel sphere

3.2 *Influence of Eccentricity*

The falling point of the steel sphere has a great influence on the shape of the sand jet. Three different impact experiments of eccentricity were conducted under the same conditions. Although the corona and jet could also be observed, the shape and the height of them are totally different.

Figure 4 shows the splash of 10 ms after impact test done under the same conditions but varying the eccentricity of falling point. Indeed, the dimension of the splash

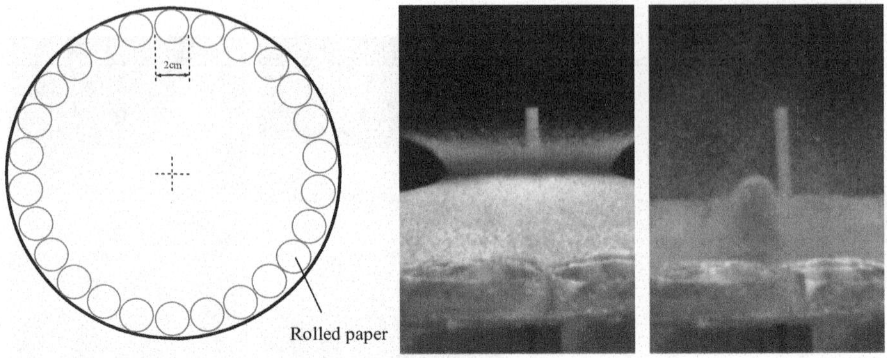

Fig. 6 Jet formation after the impact of a steel ball on quick sand with rolled paper boundary

formed by the impacting ball is less important for different eccentricity. But the eccentricity has a striking influence on the orientation of jet. Figure 5 shows that the eccentricity has a great influence on the angle of sand jet, θ. As the eccentricity increases, the angle of jet exhibits a linear decreasing trend. One possible explanation for this observation is that both air flow and surface wave behave differently when the sphere penetrates into the sand bed, forming different angle of cavity. The origin of the formation of the surface wave is the collision between the sphere and sand bed. Under eccentric condition, the surface wave first reaches the near lateral boundary and then reverberates, which produces a superposition of wave, strong enough to change the shape of jet. Therefore, changing the eccentricity of falling point leads to different jet formation characteristics.

3.3 *Influence of Lateral Boundary*

A 23×20 cm sheet of tissue paper was used to make 2 cm in diameter rolled paper around the cylinder (inside). Then a 0.4-mm mesh sieve is used to sieve sand into the container carefully. After flattening the sand surface, a steel sphere is dropt onto the sand with the soft lateral boundary. We found that varying the lateral boundary of the system indeed reduces the jet height. However, we found that the changing of lateral boundary pales in the size of the corona.

Figure 6 shows, as the side boundary is varied, a pronounced three-stage structure emerges, consisting of an initial splash, followed by a violent eruption and then a small thicker jet. Since the sand is loose and the rolled paper is soft, when the steel ball penetrates the sand bed, with the function of surface wave and gas pressure, the sand and rolled paper are compacted vertically and horizontally, respectively. These effects create a violent eruption after the corona collapses.

Because the ejection shape was very sensitive to the gas pressure and volume in the cavity [13], the left air after the previous eruption in the cavity is released

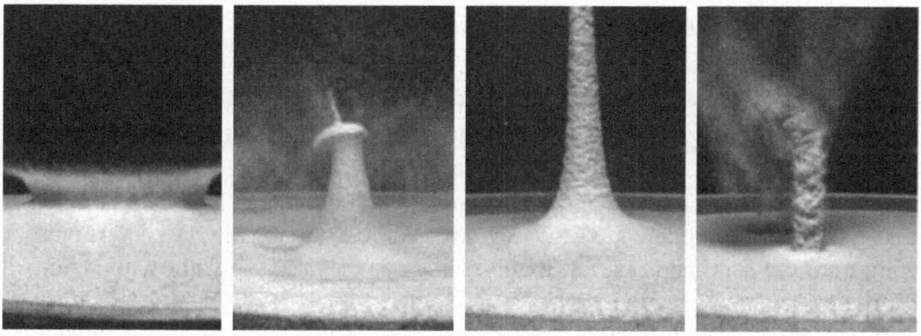

$t = 12$ ms $t = 140$ ms $t = 250$ ms $t = 460$ ms

Fig. 7 Jet formation after the impact of a steel ball on quick sand with 0.063-mm-mesh bronze sieve bottom

along with a small and thick base jet. In contrast, acrylic lateral boundary did not create eruption before jet formation and generate higher collimated jet. The eruption facilitates the outgassing and sedimentation. Despite these differences, the boundary did not affect significantly the maximum jet diameter, D_{jet}, which was found to be proportional to the steel sphere diameter.

3.4 Influence of Bottom Boundary

In order to investigate the effect of gas permeability of the bottom boundary on the jet shape, a 0.063-mm-mesh bronze sieve was used as bottom boundary instead of acrylic sheet. For the sealed bottom, following the collapse of the collimated jet, a granular eruption is seen at the position of impact, resembling a volcano, violently erasing the central peak. But for the sieve bottom, a grain jet emerges vertically and falls under gravity forming a central uplift in the crater without violent eruption.

Figure 7 shows snapshots taken from the video highlighting the main characteristics of the impact process: first, the ball vanishes in the quick sand and a corona splash is created. The corona collapses and then, as the gas is released, the surface sinks, producing two-part jet (a thin tip and a thick base topped with a cap). While the upper part of the jet is still going upwards, in the lower part resembling the surface wave driven Rayleigh instability of a water jet [6], even though there is no surface tension in quick sand. Following the collapse of collimated jet of grains under gravity, a central uplift emerges. It is important to note that no violent explosion formed after the jet collapse, in clear contrast to the violent eruption resulting from the sealed bottom.

3.5 The Distribution of Air Pressure

In order to determine the air pressure at different depth z, acrylic tubes filled with 2 cm length colored water were connected to the side boundary and bottom boundary of sand bed (Fig. 8a). The space d between the tubes below the top surface is 4 cm, the initial void length l between sand bed and colored water is 7 cm, and the radius r of perforated bottom plate is 10 cm (Fig. 8b). We let a steel ball fall from a constant height onto the sand and observe the dynamics of the colored water with a camera (29 frames per second).

As the ball vanished in the sand, the colored water starts to move outside at different velocity under the increasing air pressure. When the jet collapses and the air pressure stops increasing, the colored water keeps stable. It is driven by the air pressure at different depth z. According to the Newton's second law, for the colored water, $F=d(mv)/dt$. Here, F is the net force applied to the colored water. If we neglect the friction resistance between the colored water and acrylic tube, the inner air pressure p_i is determined, $(p_i - p_0)S=d\,(mv)/dt$. Here, p_0 is the atmospheric pressure, S is the inner sectional area of tube, m is the mass of colored water, v is the instantaneous velocity, and t is time intervals of two frames.

Figure 9a shows the increment of void volume in the acrylic tube as a function of the time for different depth. All of the experimental points are well described by the same time interval, 34.48 ms, from the impact of a steel ball to the collapse of the jet, which indicates that only v is the dominant parameter for our experiments in the studied range. Figure 9a shows that the dependence on v suggests a nonlinear relation of the increment of void volume $\triangle V$ with time t. As the depth of sand increases, the void volume in the acrylic tube increases along the side boundary, indicating that the air pressure increases from the top to the bottom. In contrast to the lateral air pressure, the bottom air pressure seems to increase slowly, even though the steel ball is near to the perforated bottom. We speculate about the difference of air permeability of sand

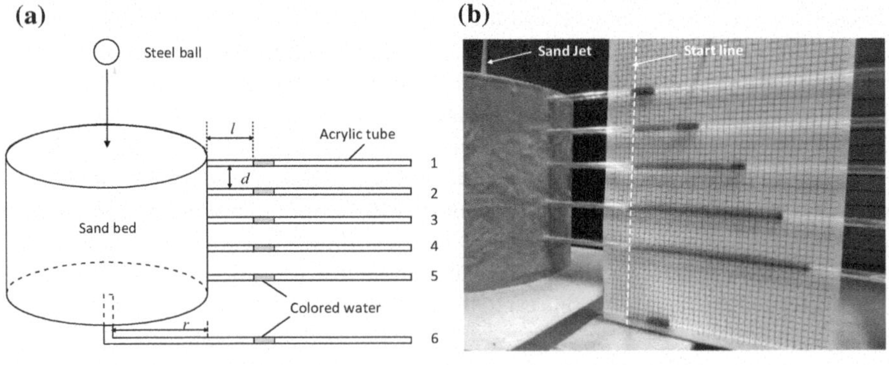

Fig. 8 **a** Schematic view of the experimental setup. **b** Different air pressure produced by steel impact

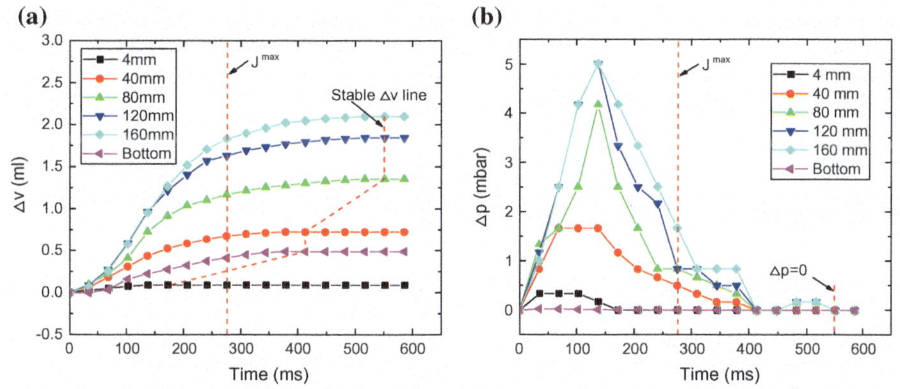

Fig. 9 **a** The horizontal increments of void volume as a function of time. **b** The horizontal increments of air pressure as a function of time

bed between the vertical direction and the horizontal layer. Only for the 1st tube, $\triangle V$ become stable before the jet reaches the maximum height. For the 2nd and the bottom tube, $\triangle V$ become stable earlier than the 3rd, 4th and 5th tube, which become stable before the collapse of the jet.

Figure 9b shows the increment of air pressure in acrylic tube of different depth. In the process of corona splash and sand jet, the air pressures along the lateral boundary increase very rapidly at first, and then decrease rapidly. After the jet reaches the maximum height, the air pressures start to level off to become asymptotic towards the x-axis.

4 Conclusions

The formation of sand jet is very complex, which depends on many influent parameters. We found that the eccentricity, lateral boundary material and gas permeability of bottom boundary play an important role in the formation of a granular jet in loose, fine sand. We have shown that the jet was inclined at eccentric impact condition because the ball did not penetrate vertically. With soft lateral boundary and sieve bottom boundary, the sand jet is less violent. After measuring the gas pressure, we found that the effect of the eccentricity, lateral and bottom boundary on the intensity of the jet is indeed closely related to the dynamic coupling between gas and granular motion. Our experiment showed that jet formation was strongly influenced by the boundaries on the cavity, presumably because the air flow around the moving ball partially levitates the grains and effectively increases the void volume and pressure in sand bed.

Acknowledgements This work was partly financed by the National Natural Science Foundation of China through Grant No. 41772304. The authors express sincere appreciation to Mr. Martin Grogger for his cooperation in performing the jet formation test. We are grateful to G. Idinger and M. A. Cabrera for helpful discussions.

References

1. Hergarten, S., Kenkmann, T.: The number of impact craters on earth: any room for further discoveries? Earth Planet Sci. Lett. **425**, 187–192 (2015)
2. Soderblom, J.M., et al.: The fractured moon: production and saturation of porosity in the lunar highlands from impact cratering. Geophys. Res. Lett. **42**(17), 6939–6944 (2015)
3. Williams, N.R., et al.: Evidence for an explosive origin of central pit craters on Mars. Icarus **252**, 175–185 (2015)
4. D'Incecco, P., et al.: Shallow crustal composition of Mercury as revealed by spectral properties and geological units of two impact craters. Planet. Space Sci. **119**, 250–263 (2015)
5. Thoroddsen, S.T., Shen, A.Q.: Granular jets. Phys. Fluids **13**(1), 4–6 (2001)
6. Lohse, D., et al.: Impact on soft sand: void collapse and jet formation. Phys. Rev. Lett. **93**(19), 1–4 (2004)
7. Lohse, D.: Bubble puzzles. Phys. Today **56**(2), 36–42 (2003)
8. Zhao, R., et al.: Granular impact cratering by liquid drops: understanding raindrop imprints through an analogy to asteroid strikes. Proc. Natl. Acad. Sci. **112**(2), 342–347 (2015)
9. Oguz, H.N., Prosperetti, A.: Bubble entrainment by the impact of drops on liquid surfaces. J. Fluid Mech. **219**, 143–179 (1990)
10. Duez, C., et al.: Making a splash with water repellency. Nat. Phys. **3**(3), 180–183 (2007)
11. Gilbarg, D., Anderson, R.A.: Influence of atmospheric pressure on the phenomena accompanying the entry of spheres into water. J. Appl. Phys. **19**(2), 127–139 (1948)
12. Lohse, D., et al.: Granular physics: creating a dry variety of quicksand. Nature **432**(7018), 689–690 (2004)
13. Loranca-Ramos, F.E., et al.: Craters and granular jets generated by underground cavity collapse. Phys. Rev. Lett. **115**(2), 028001 (2015)

Experimental and Numerical Analysis of Stick-Slip Instability of Granular Materials

Deshan Cui, Wei Wu, Wei Xiang, Qiong Chen and Shun Wang

Abstract The paper presents a comparison between experimental observations of stick-slip behaviours of glass beads reproduced in laboratory and numerical simulation obtained by two-dimensional discrete element model able to reproduce the stick-slip dynamics. A triaxial experimental set-up has been used to investigate the stick-slip response of granular matter under quasi-static loading. On the other hand the Particle Flow Code (PFC2D) has been used in this article that performs numerical simulation of stick-slip behaviour of granular matter. The model is used to simulate biaxial compression experiment on dry specimen of glass beads. Once the results are well in agreement with laboratory data, a parametric study is performed to investigate the evolution of the model's behaviour due to varying parameters such as vertical strain rate, coefficient of friction and particle stiffness. The results show an increase in critical state strength in the models with higher vertical strain rate, bigger coefficient of friction and higher particle stiffness. A new algorithm is developed to record the sliding work and sliding velocity induced by seismic events in the sample.

1 Introduction

Stick-slip, recognized as a mechanism for shallow-focus earthquakes, often accompanies frictional sliding in laboratory experiment with geological materials [1], such as glass beads and sands. However, the stick-slip event is yet to be well

D. Cui (✉) · W. Xiang · Q. Chen
Faculty of Engineering, China University of Geosciences, Wuhan, China
e-mail: cuideshan@cug.edu.cn

W. Wu · S. Wang
Institute of Geotechnical Engineering, University of Natural Resources and Life Sciences, Vienna, Austria
e-mail: wei.wu@boku.ac.at

W. Wu (ed.), *Recent Advances in Geotechnical Research*,
Springer Series in Geomechanics and Geoengineering,
https://doi.org/10.1007/978-3-319-89671-7_4

understood despite an extensive effort. Various aspects of stick-slip instability of granular matter have been already studied by laboratory experiments [2–7].

Adjemian [2] presented the study of stick-slip phenomena in dry glass beads (diameters d: 0.2 ± 0.05 mm, 0.75 ± 0.1 mm, or 3 mm) subjected to triaxial compression under small lateral confinement, <60 kPa. Roussel [8] extended the previous results in the confining stress of 100–400 kPa and found stick-slip behaviour was still observed. Doanh [3] studied the dependence of confining stress and axial strain rate of granular matter on stick-slip behaviours through triaxial laboratory experiment with confining stress of 50–400 kPa. At the same time, numerical simulations have played an increasing role to resolve the significant microscopic mechanism, such as discrete element models (DEM) [9–13]. In a biaxial shear test, Johnson et al. [14] found that above ~3 MPa load, acoustic emission and shear micro failure precursors exhibited an exponential increase in rate of occurrence, culminating in stick-slip failure.

To our knowledge, no numerical approach has been developed that is capable of accurately modeling the factors controlling the stick-slip. Compared to finite element/difference models [15], the discrete element model (DEM) has the capability to simulate geometrical heterogeneity by different particles size distribution as well as samples size. David [13] studied the mechanical response of frictional granular materials under quasi-static loading by means of two- and three- dimensional discrete element methods, finding that the decay to the critical steady state flow regime is much slower in the 3D situation than in the 2D systems.

Traditionally, the approach for modeling granular matter with PFC has been to consider friction. This approach has been successfully used in modeling the frictional problems where the frictional resistances are believed to have a correlation with normal force. However, in the present research, new routines have been developed for recording sliding work and sliding velocity, considering not only friction, but also vertical strain rate and particles stiffness. The results have been compared with the experimental data. Once the numerical results compared well with real laboratory data, a parametric study was performed to study the effect of various factors on a biaxial stick-slip behaviour.

2 Theory of Stick-Slip

Considering a simple spring-slider model with spring stiffness *k* obeying the rate-state variable friction law, as the applied driving force to the block is increased, the resistive frictional force is also increased until it reaches the maximum static frictional resistance or the ratio of shear stress to normal stress on the surface reaches the static friction coefficient. The frictional resistance drops to lower force value known as kinetic friction, the spring unloads and the block starts to move. Once the applied driving force is less than the kinetic friction force, the block decelerates until the final stop. Then there has to be a healing mechanism for friction to regain its static value. The process of periodic stops and slips is named "stick-slip", in conjunction with the Reid mechanism, recognized as one possible mechanism for shallow focus earthquakes [1].

Rabinowicz [16] proposed a critical slip distance, which he found to be approximately equal to the mean diameter of contact junctions. A slip instability will appear when $(\mu_s\text{-}\mu_d)F_n/D_c > k$, where μ_s and μ_d are static and dynamic coefficient of friction, respectively, F_n is the normal force on the block, D_c is critical slip distance, and k is the stiffness of spring.

It is obvious that the critical slip instability depends on the applied stress, critical slip distance, vertical strain rate, stiffness of the system and the surface roughness of glass beads in the laboratory. The major slip events in the triaxial compression test are believed to be the result of stress drop on the contact conjunction and consequently release of energy. It is necessary to record the sliding work and sliding velocity of particles under different compressive processes.

3 Observation of Stick-Slip Behaviours

The experimental set-up is composed of a steel frame with vertical force transducer between the frame and specimen (Fig. 1). The model was loaded in vertical direction by increasing the vertical strain rate of 0.02 mm/min. The confining stress was kept constant while the specimen was shearing.

Under triaxial compression test with confining pressure of 600 kPa (Fig. 2), the response of the granular material is firstly characterized by elastic behaviour (E) for very small deformation (less than 1% of vertical strain), and then exhibits plastic yield and critical state strength (q^{max}) at large strain. The onset ($\approx$2.5% of vertical

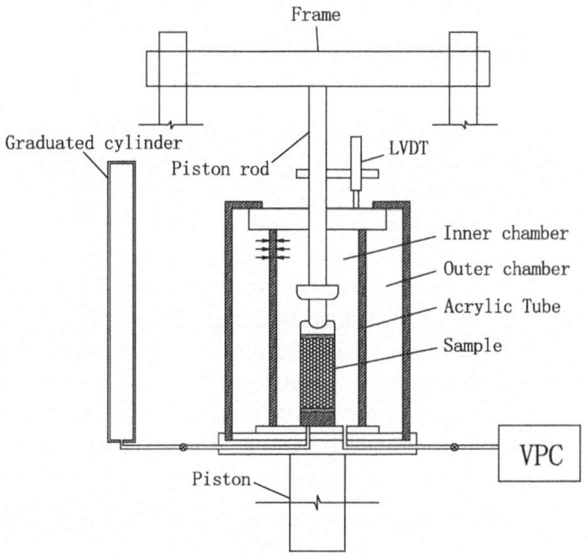

Fig. 1 The schematic of experimental set-up

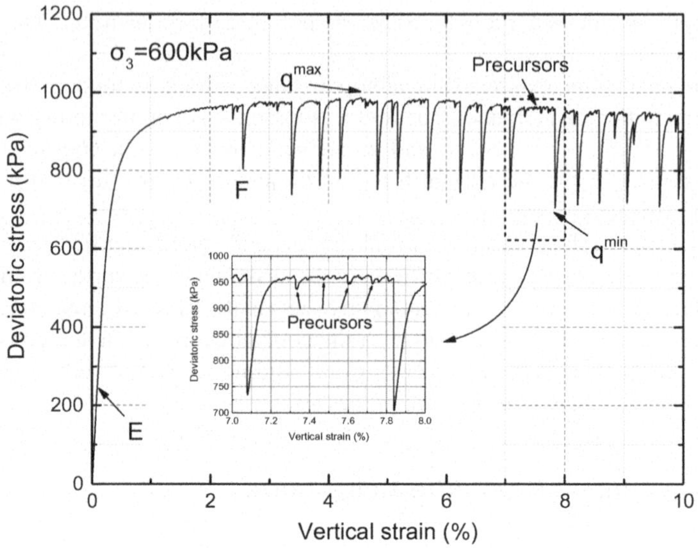

Fig. 2 Deviator stress versus vertical strain for dry granular matter, with lateral confining stress $\sigma_3 = 600$ kPa, and constant strain rate of 0.02 mm/min

strain) of instability was observed at confining pressure of 600 kPa with granular material. At this point (F), the stress-strain curve would experience several small falls and rises, and then bigger stress drop to q^{min}. Precursors or small scale deviator drops can occur in the strain softening regime before the deviatoric stress drops to minimum q^{min}.

4 Numerical Model Results

In this research, the Particle Flow Code (PFC2D) is used to model the triaxial compression test. PFC2D is an explicit implementation software of the distinct element method (DEM) developed by Itasca. In this code, it is assumed that the particles are ball and rigid. The overall constitutive behaviour of rigid particles is simulated in PFC2D by stiffness model, which gives an elastic relation between the contact force and relative displacement in the normal and shear directions considering the contact plan. In PFC, due to small time-step, a disturbance cannot propagate to farther than neighbor particles, and therefore velocity and acceleration can be considered constant during stepping. Once applying a force to the model, integrating the Law of Motion using the time step, velocity and displacement are calculated that result in the updated position of each particle. The updated positions are then used by the Force-Displacement Law to calculate the new force. This process of cycling is continued until a predefined criterion (e.g. maximum vertical strain = 10%) is met.

4.1 Algorithm for Recording Sliding Work and Sliding Velocity

The reason for slip event can be explained by using the simple static-dynamic friction law. According to State and Rate Law, the rate of decrease in frictional resistance during slip will always be faster than the rate of decrease in driving force, and thus every slip will be seismic. In biaxial compression test, assuming the stiffness of the surrounding walls is constant, vertical strain rate, confining stress, coefficient of friction and stiffness of particles are the only controlling factors for the onset of instabilities. The numerical representation of the biaxial sample is composed of all the contact points. Once a critical slip (in present research, the critical slip force is 500 kN and the critical slip work is 1 J) occurs at any of these contacts, a slip event is created and the ball is colored in red.

4.2 Numerical Simulation Results

In order to compare the experimental results and DEM simulation results, we first assure that time-step used for integration is small enough. Second, we will compare the parameters like confining pressure, coefficient of friction, vertical strain rate and particle stiffness. The DEM simulations consider two dimensional bi-axial geometries, where the ratio of height-to-width ratio of sample is 2:1. For the initial preparation of the sample, the particles with radii randomly drawn from Gaussian distribution were initially arranged on a rectangle, which is big enough for particles not to overlap. Then the 4 side walls are compressed by imposing a small confining stress ($\approx$30 kPa) in order to achieve a homogeneous, isotropic and loose initial condition. The preparation stage is finished when the kinetic energy becomes much smaller than the potential energy stored in the contacts. Before calculation, 2 side walls were deleted and two regular sheets of particles were used to identify the boundary conditions instead of the side walls. A constant strain-controlled loading is then applied through the top and bottom wall, while keeping the other stresses constant. A discrete element model of the experiment performed in the laboratory is made by PFC2D as shown in Fig. 3.

There is no direct one-to-one correspondence between the micro-parameters of discrete element model and macro-parameters of the real granular matter. Therefore, the calibration process involves some trials and attempts of varying the micro-parameters until the deserved macro response is observed. However, in the present model, in order to eliminate the effect of possible shearing band on the stick-slip behaviour, the joint and weak surfaces are not set so that they would not control the results. The granular sample parameters were chosen so that the expected stick-slip events and critical strength would be observed. The PFC2D parameters of the model are summarized in Table 1.

In PFC2D, it is possible to monitor the deviatoric stress and vertical strain values by three different ways: particle-based, measurement sphere-based and wall-based.

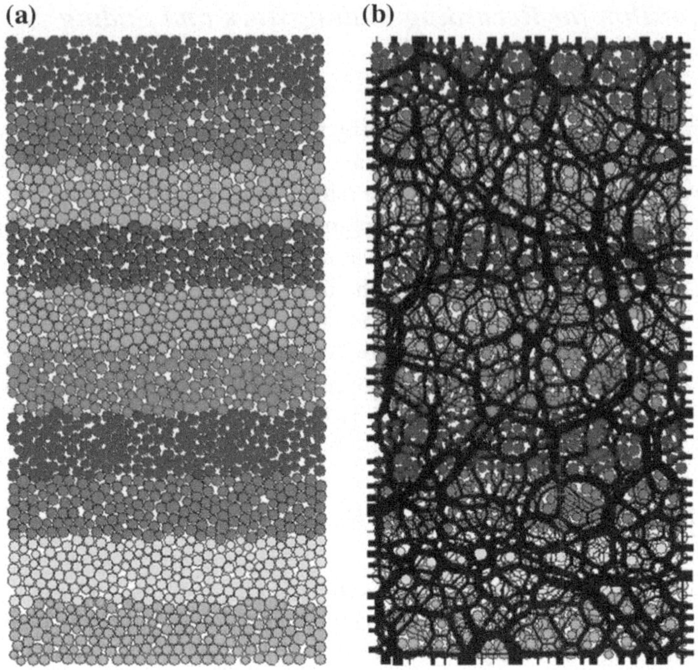

Fig. 3 Snapshot of the model systems in PFC2D with N = 1714 particles. a Arbitrary assembly of ten layers spherical particles; b Contact forces between balls and walls contacts

Table 1 Biaxial compression test parameters and material properties

Parameter	Value
Applied strain rate	0.3 mm/s
Confining stress	1e5 N/m; 2e5 N/m; 4e5 N/m
Time-step increment	1.06e-3 s
Initial height: width ratio	2:1
Number of particles	1714
Particles density	2500 kg/m^3
Smallest radius	0.75e-3 m
Largest radius	1e-3 m
Initial packing void ratio	0.175
Interparticle friction	0.15
Particle-wall friction	1.5
Maximum axial strain	10%

In order to keep the confining stress constant, the wall-based stresses are recorded. The wall-based deviatoric stress along y direction is plotted in Fig. 4a. As can be observed in this figure, there are tremendous instabilities, which exhibit drop ampli-

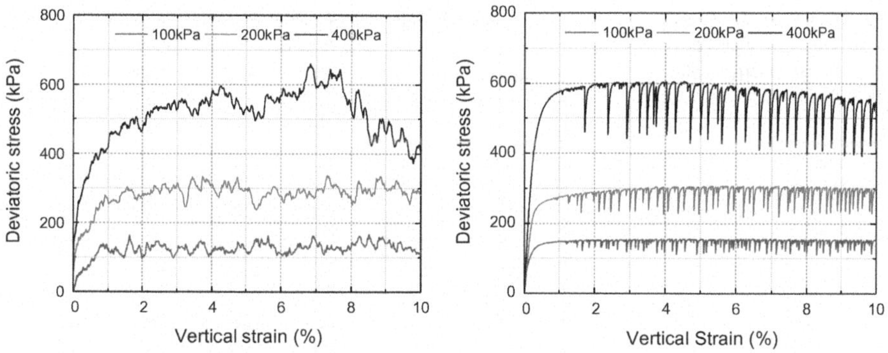

(a) Numerical results (b) Experimental results

Fig. 4 Comparison between experimental and numerical results on deviatoric stress as a function of imposed confining stress. **a** Numerical results ranging the imposed confining stress100–400 kPa and imposed velocity of 0.3 mm/s. **b** Experimental results ranging the imposed confining stress between 100 and 400 kPa and imposed velocity of 3.33E-04 mm/s

tude of deviatoric stress, just like real laboratory experiment results in Fig. 4b. In PFC2D model, the criterion of limitation is also 10% of vertical strain. Vertical displacement of the granular matter sample is calculated mathematically from the wall-based strains in the y directions.

As can be observed in Fig. 4, the stick-slips have been successfully modeled by the numerical model with the same confining stress. It has to be emphasized that the number of stick-slip events in a PFC2D model is a function of model's resolution, and therefore, it is not reasonable to compare the number of PFC2D events directly with the real number of events recorded in the experiment. A comparison between the numerical deviatoric stress and the experiment is not accurate for 2 reasons: (a) the real data belongs to the triaxial compression tests, in which the sample is cylinder; (b) the particles in this study are treated as disks, while in the real triaxial compression tests, the particles are spherical.

4.3 Parametric Study

A set of parametric studies are presented in this section to investigate the influence of various parameters on the stick-slip behaviours as well as shear events.

4.3.1 Studying the Influence of Vertical Strain Rate

The numerical results indicate that stick-slip motions involving stress drop can be sustained provided surface roughness is present. But it is not clear whether it is the

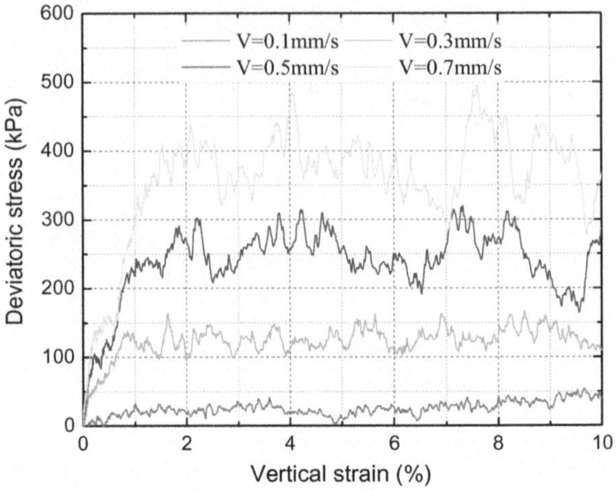

Fig. 5 Deviatoric stress versus vertical strain rate obtained from different PFC2D samples

surface roughness that alone induces stick-slip events or whether stick-slip events relate to vertical strain rate [10]. So, the influence of vertical strain rate on the stick-slip was studied. Particularly, Fig. 5 shows that the vertical strain rate plays an important role in stick-slip amplitudes. The maximum deviatoric stress increases from 54.21 to 494.96 kPa, when the strain rate increases from 0.1 to 0.7 mm/s. The variation is quite obvious, but when strain rate is too slow, the stick-slip has not developed completely. This may explain why stick-slip seems to disappear at very small strain rate, forcing stick-slip to be smaller and scarcer.

4.3.2 Studying the Influence of Coefficient of Friction

A model with the same properties as mentioned in Table 1 is repeated with different coefficient of friction (0.15, 0.30, 0.45 and 0.60). The results shown in Fig. 6 indicate an increase in deviatoric stress for bigger coefficients of friction. Higher coefficient of friction would result in greater drop amplitude of stress during stick-slip. Although the stick-slip events appear earlier in the lower coefficient of friction model, the transition state from elastic linear to instability is more obvious in the higher coefficient of friction model.

4.3.3 Studying the Influence of Normal and Shear Stiffnesses

Stick-slip instability depends on the stiffness of the system [17, 18]. A model with the same parameters as mentioned in Table 1 is repeated with varying the normal- and shear-stiffnesses of the balls (3e8, 5e8, 7e8, and 9e8 N/m). The values would represent the stiffnesses of the medium of the walls and the balls. As shown in Fig. 7, for the softest model, small stick-slip instability was observed. The peak

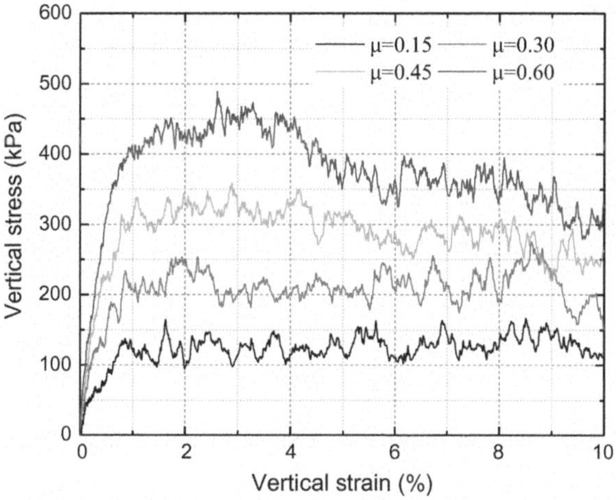

Fig. 6 Variations of stick-slip behavior in different coefficient of friction

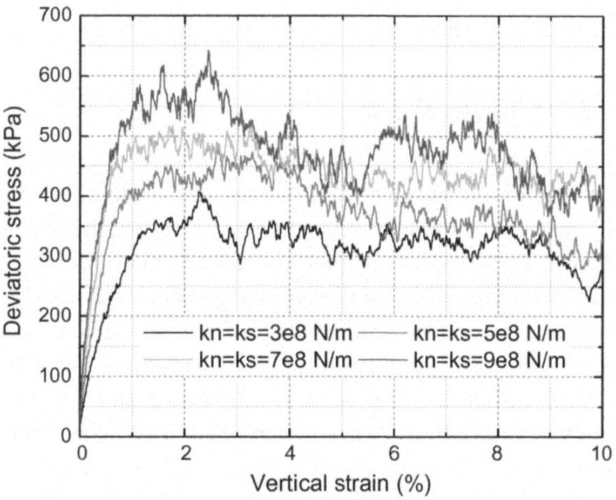

Fig. 7 Variations of stick-slip behavior in different normal and shear stiffnesses between the balls

strength, 407.84 kPa, was reached after 2.3% of vertical strain. The onset of stick-slip instabilities of softer model was much longer compared to the harder cases. For the model with the harder normal and shear stiffnesses, a softening behaviour is recorded in the compression process. With regard to the number of stick-slip events, a softer normal and shear stiffnesses has not resulted in more appearances for the same amount of vertical strain, and its only contribution has been to delay the process of stick-slips.

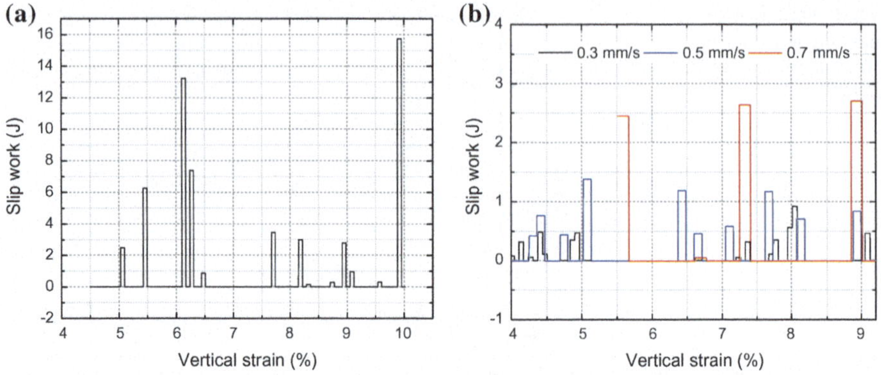

Fig. 8 Slip work of particles. **a** Slip work of ball, id = 1656; **b** Slip work of ball, id = 1689, with different vertical strain rate

5 Discussion

5.1 Sliding Work

It is found that static loading could accumulate the system work [19]. If the surfaces are in a static contact under loading for time t, then static friction coefficient increases slowly as log t [20]. Higher static friction coefficient would result in higher critical state strength, which will produce greater drop amplitude of stress during stick-slip. In order to record the slip work of single particle, a new algorithm was developed. If the different value of slip force for every particle is bigger than 500 N during stepping, then the slip work is calculated and written to a file. Figure 8a shows the slip work of ball id = 1656 when the confining stress is 200 kPa. The slip work starts to record at the vertical strain of 4.5%, then it reaches different value until to the end of the test. The maximum slip work is 15.734 J. Unfortunately, the slip phenomenon of single particle has no relationship with the system stick-slip, which indicates that stick-slip may result from lots of slips between sliding zone.

Figure 8b shows the slip work of ball id = 1689 with different vertical strain rate. The peak values of slip work are different under different vertical strain rate. With the increase in vertical strain rate (from 0.3 to 0.7 mm/s), the slip work of the granular matter increases i.e. the maximum slip work increases from 1.84 to 2.71 J. That seems to explain the stick-slip mechanism of granular matter from slip work view.

5.2 Sliding Velocity

The dynamic friction, when measured in the quasi-static sliding regime, depends on the sliding velocity [21]. In order to investigate the sliding velocity of single

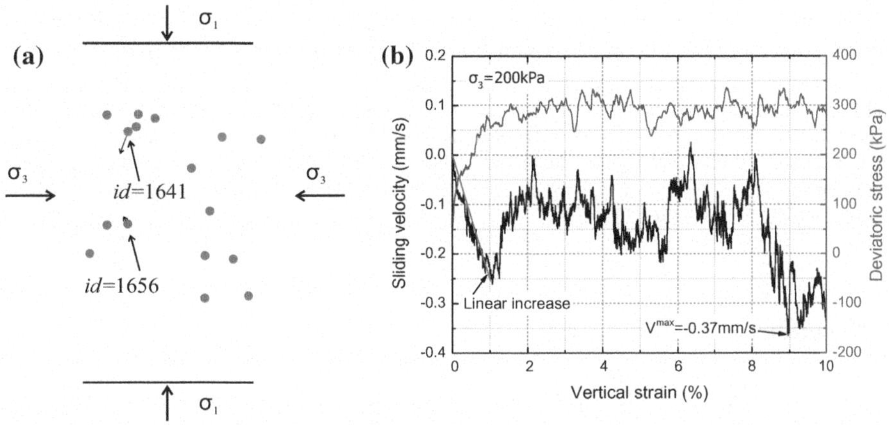

Fig. 9 Sliding velocity of granular matter. **a** Snapshot of sliding velocities of all 19 balls. **b** Y direction sliding velocity of ball (id = 1641) versus deviatoric stress of a granular simulation sample

particle, a new algorithm was used to record the sliding velocity, especially when the slip work is bigger than 1 J while stepping. As shown in Fig. 9a, the numerical simulation result can describe the sliding velocity and direction of movement of particles. There are altogether 19 particles that are colored in red. Figure 9b shows the Y direction sliding velocity first exhibits a linear increasing trend until the 1% of vertical strain, and then several small falls and rises (1 ~ 8% of vertical strain). After that, the Y direction velocity starts to increase sharply until the maximum sliding velocity, 0.37 mm/s. This important observation shows that the dynamic characters of slip instability are not only correlated with sliding of single particle.

6 Conclusions

Stick-slip behaviours of granular matter have been investigated both by experimental tests and numerical simulations. The tests performed by the experimental set-up highlight the complex stick-slip phenomena arising when granular matters are in confining state. The same system with respect to different parameters, such as imposed vertical strain rate, coefficient of friction and stiffness was established in PFC2D to simulate stick-slip instability.

From the experimental data obtained by triaxial test on glass beads, the behaviours from stable sliding to macroscopic stick-slip with constant velocity have been observed. The numerical simulations allow for investigating the other key parameters.

In future works experimental parametrical analyses will be performed in order to investigate the role of other key parameters such as surface roughness of glass beads and stiffness of the experimental frame. At the same time a further numerical investigation will be developed to quantitatively describe the mechanism of stick-slip.

Acknowledgements The authors express sincere appreciation to M. Grogger for his cooperation in performing the laboratory triaxial tests described in the paper. The authors would like to thank Prof. T. Doanh of Ecole Nationale des Travaux Publics de l'Etat, France for providing the glass beads for this study. This work was financed by the National Natural Science Foundation of China (Grants: 41672297 and 41602313).

References

1. Brace, W.F., Byerlee, J.D.: Stick-slip as a mechanism for earthquakes. Science **153**(3739), 990–992 (1966)
2. Adjemian, F., Evesque, P.: Experimental study of stick-slip behaviour. Int. J. Numer. Anal. Met. **28**(6), 501–530 (2004)
3. Doanh, T., et al.: Stick-slip behaviour of model granular materials in drained triaxial compression. Granul. Matter. **15**(1), 1–23 (2013). https://doi.org/10.1007/s10035-012-0384-6
4. Krim, J., et al.: Stick-slip and the transition to steady sliding in a 2D granular medium and a fixed particle lattice. Pure. Appl. Geophys. **168**(12), 2259–2275 (2011)
5. Lubkin, G.B.: Experiments find hysteresis and precursors in the stick-slip friction of a granular system. Phys. Today **50**(9), 17–19 (1997)
6. Lieou, C.K.C., et al.: Stick-slip instabilities in sheared granular flow: the role of friction and acoustic vibrations. Phys. Rev. E. **92**(2) (2015)
7. Janssen, R., Verwijs, M.: Why does the world need a true triaxial tester? Part. Part. Syst. Charact. **24**(2), 108–112 (2007)
8. Roussel, L.E.: Experimental investigation of stick-slip behavior in granular materials. Louisiana State University (2005)
9. Aharonov, E., Sparks, D.: Stick-slip motion in simulated granular layers. J. Geophys. Res-Sol Ea **109**(B9) (2004)
10. Mora, P., Place, D.: Simulation of the frictional stick-slip instability. Pure. Appl. Geophys. **143**(1–3), 61–87 (1994)
11. Morgan, J.K.: Particle dynamics simulations of rate-and state-dependent frictional sliding of granular fault gouge. Pure. appl. Geophys. **161**(9–10), 1877–1891 (2004)
12. Khazaei, C.et al.: Discrete element modeling of stick-slip instability and induced microseismicity. Pure. Appl. Geophys. 1–20 (2015)
13. David, C.T., et al.: Powder flow testing with 2d and 3d biaxial and triaxial simulations. Part. Part. Syst. Charact. **24**(1), 29–33 (2007)
14. Johnson, P.A., et al.: Acoustic emission and microslip precursors to stick-slip failure in sheared granular material. Geophys. Res. Lett. **40**(21), 5627–5631 (2013)
15. Nielsen, S., Tarantola, A.: Numerical model of seismic rupture. J. Geophys. Res. **97**(B11), 15291–15295 (1992)
16. Rabinowicz, E.: The nature of the static and kinetic coefficients of friction. J. Appl. Phys. **22**(11), 1373–1379 (1951)
17. Scholz, C.H.: Earthquakes and friction laws. Nature **391**(6662), 37–42 (1998)
18. Beeler, N., et al.: The roles of time and displacement in the evolution effect in rock friction. Geophys. Res. Lett. **21**(18), 1987–1990 (1994)
19. Bi, Z., et al.: Numerical study on energy transformation in granular matter under biaxial compression. Granul Matter **13**(4), 503–510 (2011)
20. Dieteric, J.H: Time-dependent friction in rocks. J. Geophysical Res. **77**(20), 3690–3690 (1972)
21. Molnar, P., Johnson, T.: Detailed studies of frictional sliding of granite and implications for the earthquake mechanism. J. Geophys. Res. **32**, 6392–6406 (1972)

Combination of Decisive Properties of Soil in Science and Geotechnics

Werner Gerber

Abstract When determining the stability of slopes, engineers use the parameters required by the calculation model. The most important properties of soil are its friction angle, density and cohesion. If a steep slope is covered with vegetation or even a century-old forest on an embankment, the values of these properties will change, due to the climate, relief and many organisms associated with the vegetation. It is these factors that make soils as described in soil science, which examines the processes of soil formation and describes many parameters other than its physical properties. Soil's physical properties mainly apply to fine soil, with a particle size of less than 2 mm. Larger components are referred to as skeleton. In geotechnics, fractions smaller than 2 mm are described as sand, silt or clay, whereas larger components are called gravel. The differences between them mainly have to do with their quantitative characterisation: soil scientists use volume-based percentages and geotechnicians mass-based percentages. These two values cannot be compared directly, but also depend on the soil's porosity and density as well as on the density of its solid substance. This article provides practical examples of the relationships between the various parameters and highlights the major differences between results obtained from a soil science point of view and from a geotechnical perspective. The physical properties of the forest soils presented here mainly relate to their density, porosity, water content and particle size distribution in the Swiss Prealps.

W. Gerber (✉)
Swiss Federal Institute for Forest, Snow and Landscape Research WSL,
Zuercherstrasse 111, 8903 Birmensdorf, Switzerland
e-mail: werner.gerber@wsl.ch

W. Wu (ed.), *Recent Advances in Geotechnical Research*,
Springer Series in Geomechanics and Geoengineering,
https://doi.org/10.1007/978-3-319-89671-7_5

Fig. 1 A shallow landslide that took place in 2005 in a large forest gap in the Praettigau Valley. Photo: C. Rickli, WSL

1 Introduction

Heavy precipitation can cause landslides and floods. In Switzerland, landslides occur mainly in mountainous regions, whereas floods cause major damage in the flatter parts of Switzerland's Central Plateau. The total damage caused by these two natural hazards is increasing in the long term by CHF 350 million per year. This increase reflects individual years associated with very large-scale events. For example, damage caused in August 2005 alone totaled roughly CHF 3 billion [1]. Damage caused by landslides accounts for somewhere between around 3% of this amount. Nevertheless, approximately 5000 landslides of various sizes were recorded in connection with this event alone [2]. If future damage caused by landslides is to be reduced, our knowledge of their causes and the processes involved needs to improve. The Swiss Federal Institute for Forest, Snow and Landscape Research (WSL) has studied various events, documenting and analysing numerous shallow landslides (see Fig. 1). This work involved collating geological, geomorphological and hydrological data. Among other things, these data are to be used to calculate slope stability with the help of simulation programs. In this connection, the main parameters are density, porosity, water content, friction angle and cohesion.

Table 1 Classification of loose material in the areas surveyed in 1997, 2002 and 2005

Area	Sachseln	Appenzell	Napf	Entlebuch	Napf	Praettigau	Total
Event	1997	2002	2002	2005	2005	2005	
Gravels	84	2	20	17	17	23	167
Sands	0	59	16	6	5	1	87
Silts, clays	141	19	10	20	10	23	223
Total	225	80	46	43	36	47	477

The data from field recordings should also be compared with data from other sources, such as geotechnical and soil science publications. This may improve our quantification of the values of individual parameters. One objective is to identify maximum and minimum limit values and compare them with one another. These are the questions we need to answer:

- In which soils do most shallow landslides occur?
- Can these soils be assigned to individual grain size classes?
- Do individual grain size classes also have uniform densities?
- Can uniform density values be expected to be accompanied by a uniform water content?
- Are the densities cited in soil science publications comparable with geotechnical data?
- Can the effective friction angle be extrapolated from density data?

2 Field Analyses of Landslides

2.1 *Classification of the Soil Material*

Numerous landslides were documented and analysed in connection with the major flood event involving heavy precipitation mentioned in the introduction. Landslides had previously been documented in connection with earlier events, from 1997 and 2002 [3, 4]. For 477 of these 522 documented landslides [5], there are data sets on the classification of soil material (see Table 1).

Almost half the landslides (223, or 47%) involved fine-grained material (silts and clays). About one-third (167, or 35%) of the classified events were landslides involving gravelly material, and nearly a fifth (87, or 18%) involved loose, sandy material. Additional analyses performed using the Unified Soil Classification System [6] show that most landslides (132) in clayey silt (CL-ML) occurred in fine-grained materials. Where medium-grained soils were involved, almost all landslides (75) occurred in silty sand (SM). Most landslides in gravelly soils (144) involved the two groups silty gravel (GM) and silty to clayey gravel (GC-GM) (see Fig. 2). Classifying

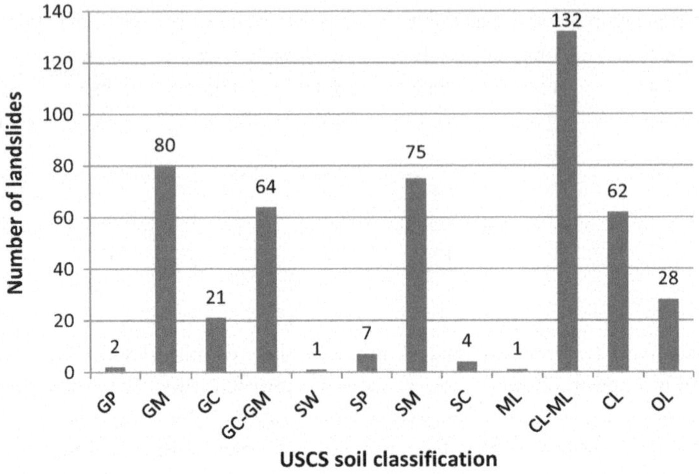

Fig. 2 Number of landslides in individual classes of gravelly, sandy, silty and clayey soils with a total of 477th

Table 2 Depth of sliding surfaces recorded in the areas under investigation in 1997, 2002 and 2005

Area	Sachseln	Appenzell	Napf	Entlebuch	Napf	Praettigau
Event	1997 (m)	2002	2002	2005	2005	2005
Mean	1.1	1.1	0.8	1.55	1.1	1.2
Median	1.1	1.0	0.8	1.3	1.0	1.2
First quant.	0.9	0.8	0.6	1.0	0.8	0.9

soils in this way enables the extrapolation from tables of further properties, such as the density of moist soils and water content (see the section headed Geotechnical Fundamentals).

2.2 *Depth of the Sliding Surfaces*

When measuring landslides in the areas under investigation, the depth of the sliding surface was also recorded [5]. Mean shallow landslides in the Napf region were found to be moderately thick (0.8–1.1 m). By contrast, landslides in the Entlebuch Biosphere were larger, 1.55 m on average (see Table 2).

Table 3 Key parameters of geotechnical soils

Gravels	GW	GP	GM	GW-GM	GC	GW-GC	GC-GM
w (%)	5.4	5.9	8.9	6.9	11.4	6.5	8.4
ρ_f (kg/m^3)	2100	2100	2340	2360	2300	2300	2370
ϕ' (°)	40	38	35	36	31.4	34	33
Sands	SW	SP	SM	SW-SM	SC	SW-SC	SC-SM
w (%)	13.0	13.2	18.8	10.8	14.8	12.6	13.7
ρ_f (kg/m^3)	1950	1940	2070	2030	2210	2000	2230
ϕ' (°)	38	36	33.6	34.4	34.8	32	35.3
Silts and clays		ML	CL-ML	CL	CM	CH	OL
w (%)		28.5	19.8	19.4	25.0	34.5	46.3
ρ_f (kg/m^3)		1970	2110	2120	2020	1880	1750
ϕ‘ (°)		33.9	32.7	30.7	25	24	27

w Water content, ρ_f Moisture density, ϕ' Effective friction Angle

3 Geotechnical Fundamentals

3.1 *Water Content and Density*

For simple construction projects, general information about soil properties can be found in Swiss standard SN 670010 [7]. Building ground is divided into gravels, sands, silts and clays. Gravels and sands are each subdivided into 7 classes; fine-grained soils are subdivided into 6. The most important properties, like moisture density ρ_f, water content w and effective friction angle ϕ' are listed in Table 3. Only silts and clays with a water content of less than 50% are listed (field analyses found no soil types with a water content higher than 50%).

Naturally, the moisture density depends on the water content, which fluctuates over time in line with precipitation. However, to enable density comparisons, moisture density and water content values are converted to dry densities ρ_d. This can be done, provided that the density of the solid substance ρ_s is determined in the laboratory or defined in general terms. The density of the solid substance in Switzerland is assumed to have a value of around 2700 kg/m^3.

The dry density is calculated using formula (1), whereby the water content is not given in per cent, but as the number (w/100).

$$\rho_d = \frac{\rho_f}{(1 + w)} \tag{1}$$

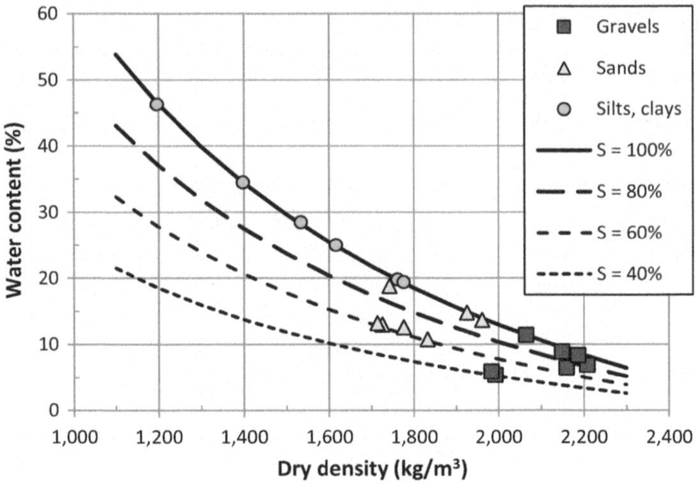

Fig. 3 Dry density values calculated on the basis of moisture density and natural water content at a solid substance density of 2700 kg/m3

Clear differences between the different classes of soil now emerge, with respect to both water content and dry density. Silts and clays have a dry density of between 1200 and 1780 kg/m^3, whereas sands range between 1710 and 1960 kg/m^3 and gravels occupy a narrow range of values between 1980 and 2210 kg/m^3. The explanation for this lies in the fact that different soils have varying porosities. Porosity n is calculated using formula (2) and shows how much air and water a specified soil volume contains. Porosity is expressed as a number, though informal circles also talk of percentages ($v\%$).

$$n = 1 - \frac{\rho_d}{\rho_s} \tag{2}$$

At the fixed solid substance density of 2700 kg/m^3, gravel porosity ranges from 0.18 to 0.27, sand porosity from 0.27 to 0.37 and fine-grained soils from 0.34 to 0.56. Porosity will vary slightly with changes in the density of the solid substance ρ_s in a range of 2600 to 2750 kg/m^3. For example, porosity at a dry density of 1600 kg/m^3 varies within a narrow range of between 0.38 and 0.42. For fine-grained soils, the range of variation is even smaller, whereas for gravels it is slightly broader. To analyze the porosity, first we had the determine the dry density with Fig. (3). With this value we analyze the porosity in Fig. (4).

Porosity limits the soil's water content. If all air pores fill with water, this represents 100% saturation, but the level of saturation will fall if some pores in the soil contain air as well as water. The natural state is primarily characterised by partial saturation in well and poorly graded gravels (GW and GP). For example, the saturation level of the gravels in question here is only around 40%. The situation for sands is similar, with

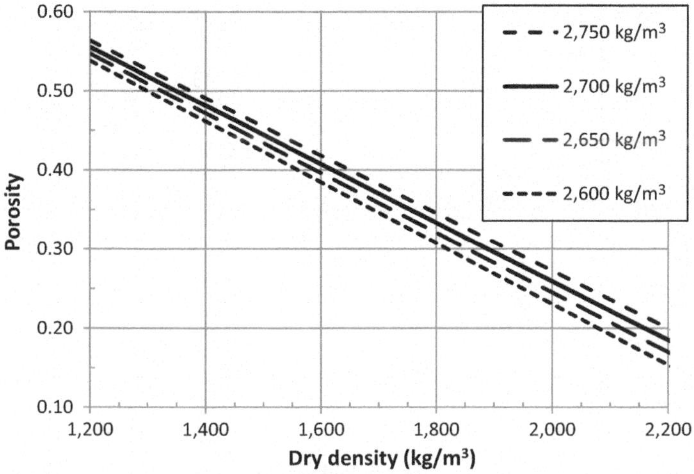

Fig. 4 Porosity at various dry density values and solid substance values

well and poorly graded sands saturated to a level of approximately 60%. According to the figures cited in the standard, all silts and clays are presented with a saturation level of 100% (see Fig. 1).

Different porosities impose an upper limit on the water content *w* (%). The maximum water content can be calculated from the dry density using formula (3). In this context, the density of water is taken to equal $\rho_w = 1000$ kg/m^3.

$$w\_max = \frac{n \cdot \rho_w}{\rho_d} \tag{3}$$

The maximum water content of gravels with dry densities of between 2000 and 2200 kg/m^3 is between 8 and 13%, whereby the lower water content value applies to higher-density soil. The maximum water content of sands with dry densities of 1700–2000 kg/m^3 is between 13 and 22%. The maximum water content of silts and clays ranges very widely, from 18 to 50%. The water content of soils with dry densities of less than 1100 kg/m^3 can exceed 50% (see Fig. 3).

3.2 Friction Angle

Another key parameter for defining the properties of soil is the effective friction angle ϕ'. The standard [7] also gives the friction angle for normally consolidated soils. Within individual classes of soil, the friction angle varies to a relatively large extent, namely by ±4°–5°. The values for silts and clays range between 17° and 28° and at coarser particle sizes between 32° and 40°. These values can be higher for denser soils or correspondingly lower for more loosely packed soils (see Fig. 5).

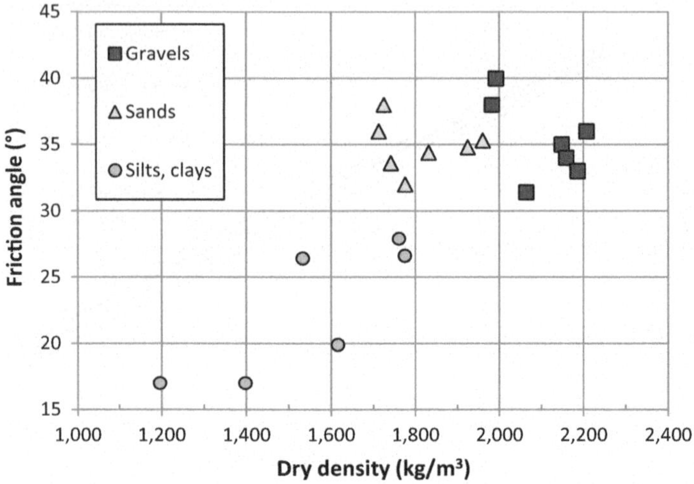

Fig. 5 Friction angles of various classes of soil with different dry densities

4 Pedological Fundamentals

Soil scientists study the uppermost level of the Earth's crust, more specifically analysing the impact of climate, relief and organisms on parent rock over time. The results for forest soils in Switzerland are divided up into the following regions: Jura, the Swiss plateau, the Prealps, Alps and southern side of the Alps [9].

4.1 Dry Densities

Soil classifications in the field describe their density, including their pressure on a sample and the penetration resistance of a knife in a sample. The result is a range of classes from very loose to extremely dense [8]. The laboratory method involves taking a set volume of soil from the field in a cylinder and weighing it. This indicates the soil's moisture density ρ_f in the natural state. Once the sample has dried, it is re-weighed to give the dry density ρ_d. When determining dry density, the volume of the sample (for example arising from the shrinkage of the sample during drying) remains unchanged. This methodology is in line with geotechnical methods, and the data in similar soil conditions should in principle be comparable.

All the landslides investigated by the field analyses occurred in the Prealps, except for one that occurred in the Praettigau Valley region in 2005. Densities of 9 gley and pseudo-gley soils from this area were used for comparison with the geotechnical fundamentals. The dry densities cited in Zimmermann et al. [9] and soil skeleton content at different depths are presented. On average, 2 data pairs were taken per soil

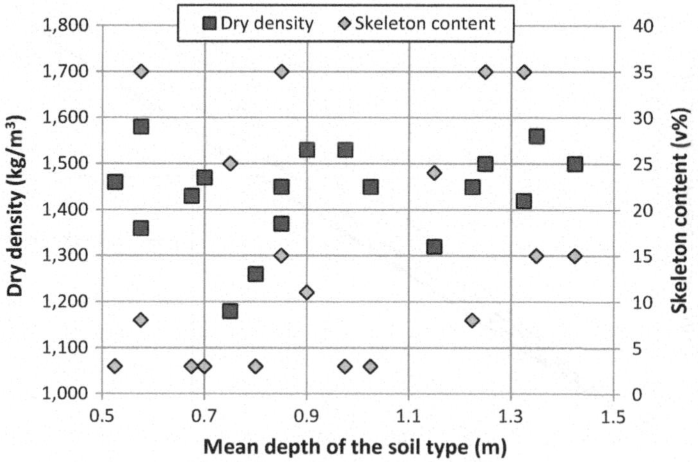

Fig. 6 Dry density and skeleton content at various depths of gley soils

layer. The densities found ranged from 1180 to 1580 kg/m^3 for skeleton contents of between 3 and 35% by volume (see Fig. 6). Instead of determining accurate skeleton content values at every depth, individual values were assigned to different classes.

4.2 *Skeleton Content*

The methods for determining skeleton content are described in Walthert et al. [8]. The solid part of the soil is subdivided into two components: fine earth and skeleton. Fine earth is all material with a particle diameter of less than 2 mm; larger components qualify as skeleton. The proportion of skeleton in the soil sample is stated in per cent by volume (v). This is in contrast to the geotechnical method, in which individual components are represented as per cent by weight (g) in the particle size distribution curve (see Fig. 8).

While the proportion of skeleton is described in per cent by volume (v), the mass depends on the density of the solid substance ρ_s. Formula (4) can be used to convert the figures into per cent by volume.

$$g = \frac{v \cdot \rho_s}{\rho_d} \tag{4}$$

The figures in per cent by weight also depend on the respective soil's dry density ρ_d, but will definitely be higher than the figures in per cent by volume. For example, the values at a dry density of 1800 kg/m^3 and a solid substance density of 2700 kg/m^3 will be 1.5 times more than the figures in per cent by volume (see Fig. 7).

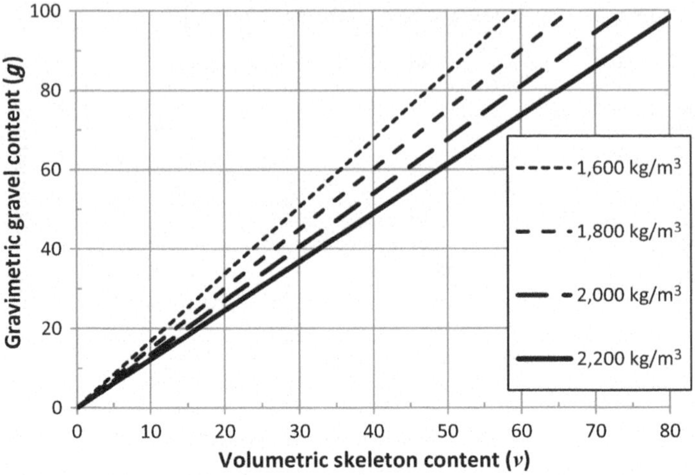

Fig. 7 Comparison of the volumetric skeleton content (soil science) and gravimetric gravel content (geotechnology)

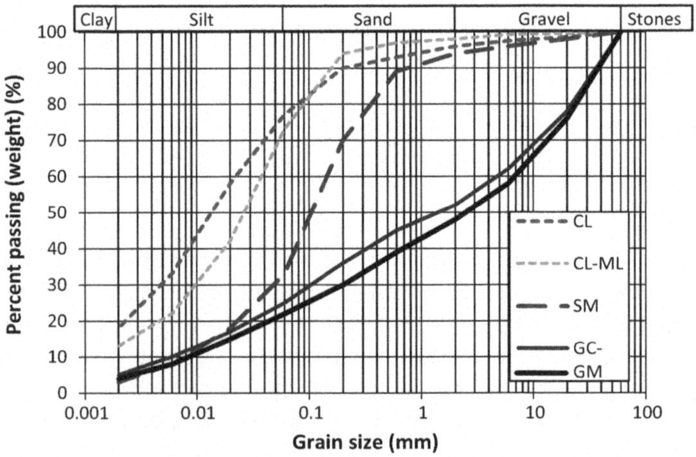

Fig. 8 Grain size distribution curves of the classes of soil most frequently involved in shallow landslides

5 Comparison and Discussion

The classifications of the soil material show that the majority of landslides occur in silty and clayey soils (see Fig. 2). Even in gravels and sands, the proportions of silt and clay fraction were key to their classification. The two most frequently occurring fine-grained soils (CL and CL-ML) contain 73–77% silt and clay (see Fig. 8).

However, it is more important for these two soils to have a dry density of 1750 kg/m^3 and a water content of 20% (see Fig. 3). At this dry density, porosity is 35% (see Fig. 4) while the indicated water content means a saturation level of 100%. Assuming a natural saturation level of 80%, all it will take is 70 mm of precipitation to saturate the soil down to a depth of 1 m.

In their natural state, the two gravelly soils (GM and GC-GM) have a water content of 7–8% at a dry density of 2200 kg/m^3. Saturation levels of between 83 and 95% can be calculated from these data. This being the case, it would take another 10–30 litres of water or 10–30 mm of precipitation to fully saturate the top metre of soil.

These calculations show that in some situations, relatively little water is all it takes to fully saturate the soil. In silty gravels with a high dry density significantly less water or precipitation is required than in fine-grained soils with lower dry densities to achieve full saturation.

Consequently, the density of soil material is a key factor in determining the properties of soil. In our paper, the densities of individual classes of soil are taken from tables and diagrams. However, it would be better to take moist soil samples in the field using cylinders, dry them and determine their water content and dry density in the laboratory. That way, actual values, as opposed to not generally applicable ones could be used to calculate the stability of slopes.

The same applies to determining soil material's friction angle. However, this will require a vastly greater effort. Here the actual dry density could be used to estimate the friction angle. Other methods could also be used to estimate the friction angle based on the particle size distribution curve.

The dry density values published in the basic pedological literature for gley soils (see Fig. 6) are roughly equivalent to the values for silts and clays (taken from geotechnical publications). In both cases, values ranging from 1200 to 1600 kg/m^3 are cited. However, in the geotechnical literature, the values for fine-grained soils rise as high as 1800 kg/m^3. Generally speaking, the dry densities rise as the skeleton content increases (see Fig. 3). Yet this picture does not clearly emerge from the published pedological data. Perhaps more different pedological soil types should be analysed, to confirm this trend.

The gravel content in geotechnically loose material varies from 0 to 80%. Fine-grained soils rank lowest, whereas coarse soils (gravels) score the highest value. Even well-graded gravels (GW) contain roughly 20% sand, silt and clay. Dry densities of between 1650 and 2000 kg/m^3 result in volumetric skeleton contents of between 50 and 60%. According to the geotechnical figures and the indicated conversion (using formula 3), volume shares higher than 60 mm are impossible for loose material with a particle diameter of 60 mm. Nonetheless, pedological methods do record one class with a value in excess of 75% (containing extremely skeleton-rich soil). If a soil has a skeleton content of 80% and a dry density of 2200 kg/m^3, the soil sample would no longer contain any fine earth (< 2 mm) (see Fig. 7).

Acknowledgements The Author thanks the Project "REVENUES" for supporting and funding this research: Marie Curie Actions of the European Union's Seventh Framework Programme FP7-PEOPLE-2012-IAPP (grant agreement n° 324466).

References

1. Bezzola G.R., Hegg C. (Ed.): Ereignisanalyse Hochwasser 2005, Teil 1—Prozesse, Schaeden und erste Einordnung. Swiss Federal Office for the Environment (BAFU), Federal Institute for Forest, Snow and Landscape Research (WSL), Environmental knowledge, no. 0707, 215 pages 2007
2. Bezzola, G.R., Hegg, C. (Ed.): Ereignisanalyse Hochwasser 2005 Teil 2—Analyse von Prozessen, Massnahmen und Gefahrengrundlagen. Swiss Federal Office for the Environment (BAFU), Federal Institute for Forest, Snow and Landscape Research (WSL), Environmental knowledge no. 0825, 429 pages (2008)
3. Rickli C. (Red.): Vegetationswirkungen und Rutschungen—Untersuchungen zum Einfluss der Vegetation auf oberflaechennache Rutschprozelle anhand der Unwetterereignisse in Sachseln OW am 15. August 1997. Birmensdorf, Bern; Federal Institute for Forest, Snow and Landscape Research (WSL), Swiss Agency for the Environment, Forests and Landscape (SAEFL), 97 pages (2001)
4. Rickli, C., Bucher, H.: Oberflaechennahe Rutschungen im Napfgebiet und im Gebiet Appenzell. Project report submitted to the Swiss Federal Office for Water and Geology (BWG), 96 pages (2003)
5. Rickli, C., Kamm, S., Bucher, H.: Ereignisanalyse Hochwasser 2005 Teilprojekt Flachgruendige Rutschungen. Project report submitted to the Swiss Federal Office for the Environment (BAFU), Hazard Prevention Division; 114 pages (2008)
6. VSS 2011: Geotechnische Erkundung und Untersuchung—Geotechnische Kenngroessen. Swiss standard SN 670 010. Swiss Association of Road and Traffic experts (VSS), 19 pages
7. VSS: Identifikation der Lockergesteine—Labormethoden mit Klassifikation nach USCS. Swiss standard SN 670 008a. Swiss Association of Road and Traffic experts (VSS), 16 pages (1997)
8. Walthert, L., Zimmermann, S., Blaser, P., Luster, J.L., Luescher P.: Waldboeden der Schweiz. Band 1. Grundlagen und Region Jura. Birmensdorf, Federal Institute for Forest, Snow and Landscape Research (WSL). Bern, Hep Verlag, 768 pages (2004)
9. Zimmermann, S., Luster, J., Blaser, P., Walthert, L. Luescher, P.: Waldboeden der Schweiz. Band 3. Regionen Mittelland und Voralpen. Birmensdorf, Federal Institute for Forest, Snow and Landscape Research (WSL), Bern, Hep Verlag, 848 pages (2006)

Effects of Mycorrhizal Fungi on Slope Stabilisation Functions of Plants

Frank Graf, Alexander Bast, Holger Gärtner and Anil Yildiz

Abstract Plants essentially contribute to the strength of soil and, in particular, steep slopes susceptible to erosion and shallow landslides. The corresponding functions of vegetation significantly control processes above and below ground such as interception, evapo-transpiration, soil aggregation and root reinforcement. Either way, they are all correlated with plant growth. Consequently, in order to unfold their soil stabilising potential, the plants must grow and sustainably survive. However, what sounds so obvious is anything but given under the often hostile conditions dominating on bare and steep slopes. This is exactly the point where mycorrhizal fungi come into play, known to improve the plants' ability to overcome periods governed by strongly (growth) limiting factors. Within this scope, numerous investigations have been conducted in order to understand and quantify mycorrhizal effects on different plant and soil functions related to eco-engineering and, particularly, to soil and slope stability. Results on plant growth and survival as well as on soil aggregation and slope stabilisation are presented and discussed from a mycorrhizal perspective.

F. Graf (✉) · A. Yildiz
WSL Institute for Snow and Avalanche Research SLF, 7260 Davos Dorf, Switzerland
e-mail: graf@slf.ch

A. Yildiz
e-mail: anil.yildiz@igt.baug.ethz.ch

A. Bast · H. Gärtner · A. Yildiz
Swiss Federal Institute for Forest, Snow and Landscape Research WSL, 8903 Birmensdorf, Switzerland
e-mail: alexander.bast@wsl.ch

H. Gärtner
e-mail: holger.gaertner@wsl.ch

A. Yildiz
IGT Institute for Geotechnical Engineering, ETH Zürich, 8093 Zurich, Switzerland

W. Wu (ed.), *Recent Advances in Geotechnical Research*,
Springer Series in Geomechanics and Geoengineering,
https://doi.org/10.1007/978-3-319-89671-7_6

1 Introduction

Shallow landslides have been a constant threat in mountainous regions, predominantly on steep slopes, and the probability of occurrence of this natural hazard will increase in future, driven by the predicted intensification of heavy rainstorm events [1]. In Switzerland, substantial damage on infrastructure of up to more than 100 million Swiss Francs related to one heavy rainstorm period and sporadically loss of lives have been caused by erosion and landslides in recent years [2]. The protection against as well as prevention and prediction of superficial soil failure are, however, still difficult.

The importance of healthy and intact vegetation in general and forests in particular with regard to slope stability has long been accepted. Even as far back as in ancient Greek it was taught that one has to be concerned about trees and forests in order to ensure the well-being of the Nymphs, the guardians of the woods. Without having access to databases of meteorology and natural hazards, the important regulatory functions of forests related to hydrological processes were well known [3]. Accordingly, the gods punished sinners causing harm as for example Erisichthon who felled a banned tree but rewarded benefactors as in the case of Arkan having protected the tree of a Nymph who gave him her love [4, 5].

Today, these wisdoms are common knowledge and it is well accepted that vegetation considerably contributes not only in respect of regulating water balance but generally in view of protection against natural hazards, such as avalanches, rock-fall or debris flows and, predominantly, shallow landslides and erosion. Consequently, vegetation and the application measures of eco-engineering are widely recognised. In Switzerland incentive and support for biological stabilisation is given by article 19 of the forest law [6]. With regard to superficial soil failure, the protective functions include hydrological regulation through interception and evapo-transpiration as well as soil mechanical and, to a certain extent, chemical stabilisation through root reinforcement (Fig. 1).

Interception by forest canopies accounts for 10–40% of precipitation [7]. However, values vary spatially and depend on rainfall rate as well as on type and structure of forest. Additionally, a more or less distinct seasonal dependency exists. Generally, interception tends to be higher for conifer dominated than for pure broad-leafed forests [8–12]. Yet, regardless of tree species, the interception potential is positively correlated with leaf area [13] and, therefore, strongly related to plant growth performance.

Evapo-transpiration is tightly coupled to interception and describes the process where water within a plant coming from the roots is subsequently lost as vapour through stomata in its leaves. Factors affecting evapo-transpiration include solar radiation, humidity, temperature and wind, as well as the percentage of vegetation cover, species composition and their growth stage. As far as the plant specific parameters are concerned, again, a positive correlation exists between the amount of water transport and biomass, reflected by both roots and shoots whereby leaf area is the dominating driver.

Fig. 1 Soil stabilising functions of plants, e.g. interception, evapo-transpiration, and root reinforcement

Another biomass and, in particular, root length dependent process contributing to the upward water movement is hydraulic lift, a passive dislocation of water from parts of the root system in moister soil layers into soil layers with lower water potential. Usually, considerable amounts of water are transferred from wetter (deeper) layers to the often drier near surface zone of the soil during the night. This partial rehydration of the upper soil layers provides an additional source for transpiration the following day. Lifted water may also contribute to the availability of water soluble nutrients located most plentiful in the upper soil layers and, therefore, indirectly influences survival and growth performance of the plants and associated organisms. Hydraulic lift may prolong or enhance fine-roots activity in the subsurface layers by keeping them hydrated and thus, buffer the rhizosphere organisms from effects of soil drying during persistent periods of lacking precipitation [14, 15].

Root length density does not only strongly influence hydrology, but considerably contributes to reinforcement and aggregation of soil and, consequently, to slope stability. The significant positive correlation between rooting intensity and slope stability has been frequently proven using different methods to test for soil strength, e.g. direct shear tests [16], triaxial compression tests [17], and soil aggregate stability analysis [18–21] considering different approaches and models [22–25].

All the aforementioned plant specific functions related to slope stability require plant growth and development. However, what sounds so obvious is anything but given, particularly under the often hostile conditions dominating on bare and steep slopes. At exactly this point mycorrhizal fungi, the symbiotic partners of almost all plants used in eco-engineering, come into play. Within this mutualistic partnership the fungus profits from the photosynthetic products, synthesised by the plant and, for its part, supplies the hosts with water and nutrients improving the plants' ability to overcome periods governed by strongly (growth) limiting factors [26]. Numerous studies have shown that mycorrhizal fungi improve plant growth and their survival [21, 26, 27].

Within a functional mycorrhizal symbiosis, the above ground biomass as well as the root network of the fungi's host plants increase considerably compared to non-mycorrhized plants (Fig. 2), due to a much better and more efficient nutrient supply. Read and Boyd [28] measured between 10 and 80 m of ectomycorrhizal hyphae per 1 cm of fine roots of Scots pine seedlings. These immense hyphal networks explore the soil more intensely than roots. Compared to root hairs, fungal hyphae are three to five times smaller in diameter (2–5 μm) and, consequently, have partially access to meso-pores (0.2–10 μm) that root-hairs don't. Thus, the mycelia of the host plants' mycorrhizae increase the absorption surface for water and nutrients up to 50 times [29].

Furthermore, mycorrhizal fungi affect soil aggregate stability indirectly through their host plants, particularly by accelerating the development of their root network and by serving as a distribution vector for associated micro-organisms, themselves soil stabilising alike [30–34].

In addition, the fungi directly contribute with their filamentous growth-form and the vast mycelial networks, growing far beyond the rhizosphere, to aggregate stabilisation by enmeshing loose soil particles and cement them through the production of "sticky" metabolites such as polysaccharides and hydrophobines [35–38].

Subsequently, we present and discuss results of our investigations in order to quantify mycorrhizal effects on plant growth and survival along with soil functions related to eco-engineering in general and to soil and slope stabilisation in particular.

2 Materials and Methods

2.1 Plant Survival

In six research plots, established in the subalpine forest belt in Grisons (Switzerland), the survival rates of mycorrhizal inoculated and non-inoculated plants were determined at the end of the first three growing seasons after their planting in May 2010 [20, 21]. Survival was determined for alder (*Alnus viridis* (Chaix) DC., *Alnus incana* (L.) Moench.) and willow species (*Salix purpurea* L., *Salix sp.*) as the ratio between the originally planted individuals and the still living plants after the corresponding vegetation period. The commercial inoculum "Forest", hereinafter the "commercial inoculum", includes 8 ecto- and 3 arbuscular mycorrhizal fungi [39].

In addition, the difference of survival between inoculated and non-treated dwarf willows (*Salix herbacea* L.) planted as cuttings on a graded alpine ski slope (2535 m

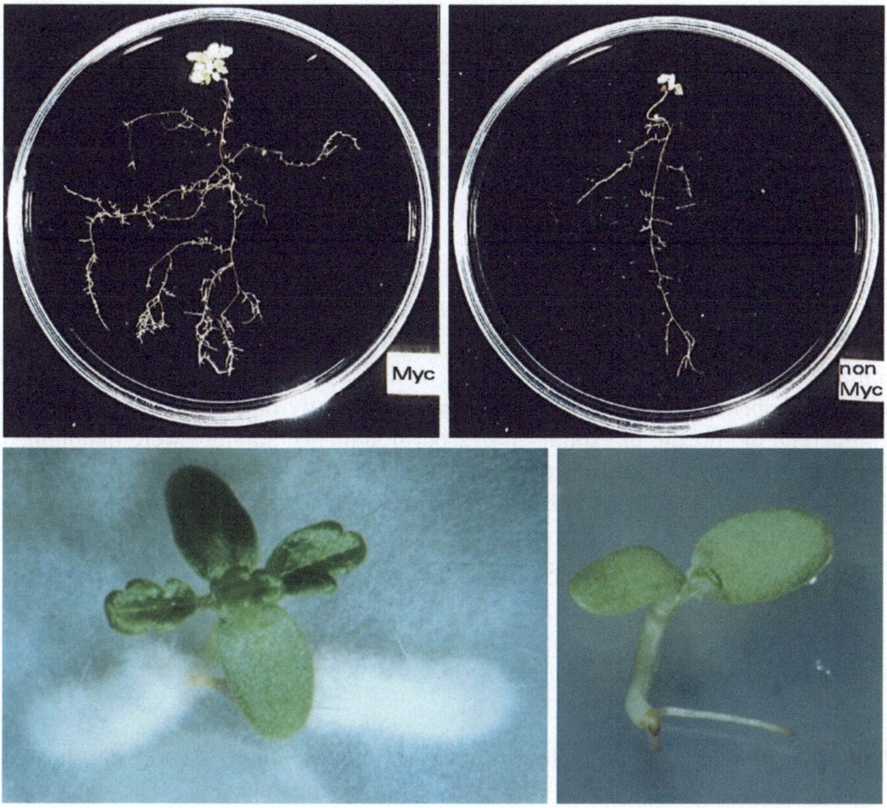

Fig. 2 Roots of a 20 weeks old *Dryas octopetala* plantlet and a 2 weeks old seedling inoculated with *Laccaria bicolor* showing a highly branched root network and the fluffy white mantle and mycelium of the fungus (left) and corresponding non-inoculated controls (right)

asl) in Grisons, Switzerland was investigated [40]. Three representatives of the genus *Laccaria* (two strains of *L. bicolor* (Maire) P.D. Orton, and *L. montana* Singer) where used as inoculi, previously isolated and cultivated from fruit-bodies of naturally grown symbiotic partners of *S. herbacea* plants in an alpine environment [41].

2.2 Plant Growth

Plant growth performance was determined from 5 months old birch seedlings (*Betula pendula* Roth) used for aggregate stability tests (cf. below). Three different treatment series with 12 individuals each were analysed, including one series of non-treated and two series of inoculated specimens. The two inoculi applied were the commercial and a single species, *Hebeloma crustuliniforme* (Bulliard) Quél., from the mycorrhiza culture collection of WSL, subsequently referred to as "specific inoculum".

For the above ground biomass, the dry weight was measured. In respect of the belowground part, root length density was determined with the software WinRhizo® [42] following a standardised protocol [43].

2.3 Soil Aggregate Stability

Samples were prepared with the fractions $\leq$10 mm of a soil type "ML" [44] from a recent landslide area in St. Antönien (GR, Switzerland). The material was tamped into PVC plastic tubes (diameter: 70 mm; height: 140 mm) aiming at a specific dry unit weight of $\approx$15 kN m^{-3}. Treatments included untreated control, soil planted with *Betula pendula* as well as planted and additionally inoculated with either the commercial or specific inoculum ($n = 12$ per treatment). Samples were maintained in a greenhouse for 20 weeks according to a standardised procedure [43].

Soil aggregate stability was determined with a wet-sieving procedure (mesh opening 20 mm) and defined as ratio between the dry weight of the components above the sieve (aggregates > 20 mm) and the sum of all components [25, 43].

2.4 Direct Shear Tests

Planted and unplanted large-scale specimens (50 cm × 50 cm × 30 cm) were sheared to failure in an Inclinable Large-scale Direct Shear Apparatus [16]. The specimens were subjected to an artificial rainfall event prior to shearing in order to ensure saturation [45]. The plant assortment included *Alnus incana* (L.) Moench, *Poa pratensis* L., and *Trifolium pratense* L. The growth period in a light, temperature, and humidity controlled climatic chamber was 6 months [16].

2.5 Statistical Analysis

Statistical inference was performed with the software R 3.3.1 [46]. Based on the Kaplan-Meier estimate [47, 48], the difference between the survival curves of the two treated sites was tested by using the Mantel-Haenszel test [49] of the R-package "survival".

Differences in soil aggregate stability and root length density were calculated using robust Kruskal-Wallis and pair-wise Wilcoxon rank sum tests. In order to test the effect of root length density on soil aggregate stability, robust simple linear regression models were fitted using "lmrob" from the R-package "robustbase". The response variable was transformed ($\sin^{-1}\sqrt{y}$) following the approach of "first aid transformations" [50].

Residual analysis was conducted in order to check the compliance of the assumptions required and the fit of the linear regression model (Appendix). The corresponding tests included: sample against theoretical quantiles (normal quantil-quantil plot), residuals against fitted values (Tukey-Anscombe plot), residuals against leverages (hat matrix, Cook's distance), and the distribution of the residuals (histogram).

3 Results

3.1 Plant Survival

Within the large-scale investigation on the influence of mycorrhizal inoculation on the performance of eco-engineering measures in a very steep catchment of a subalpine forest belt, it became obvious that survival of plants inoculated with a commercial product [39] was considerably higher compared to the non-treated control plants as shown in Fig. 3 [20, 21, 51].

The survival rates of the two alder species (*Alnus incana, A. viridis*) as well as of the two willows (*Salix purpurea, Salix sp.*) treated with the commercial inoculum performed significantly better than the corresponding controls right from the outset. Independent of treatment, the willows survived far better and obviously outperformed the alders (Fig. 3). After 3 years, more than 50 and 80%, respectively, of the mycorrhized alder and willow saplings were still alive compared to barely 20 and 50% of the corresponding controls [51].

Similar results were found for dwarf willow cuttings of *Salix herbacea* on a graded alpine ski slope. A significant increase in survival of the mycorrhized plantlets was recognised compared to non-inoculated controls.

However, the time span until the positive effect on survival of the mycorrhized plantlets was established, depended significantly on the mycorrhizal fungal species, although they all belonged to the genus *Laccaria* and two of them even to the same species (different strains). Whereas the cuttings inoculated with strain 2 of *Laccaria bicolor* yielded significant higher survival rats already after the first vegetation period this was the case only after the second year for strain 1 of *L. bicolor* and not until the end of the third year for the species *L. montana* (Fig. 4).

3.2 Plant Growth

Growth performance with regard to the above ground biomass of *Betula pendula* was significantly different for the two treatments with mycorrhizal fungi as well as compared to the non-inoculated control plants (Figs. 5 and 6).

Whereas the specific inoculum yielded no effect, the plants mycorrhized with the commercial one produced significantly higher shoot dry weights compared to both the controls and the treatment with the specific inoculum.

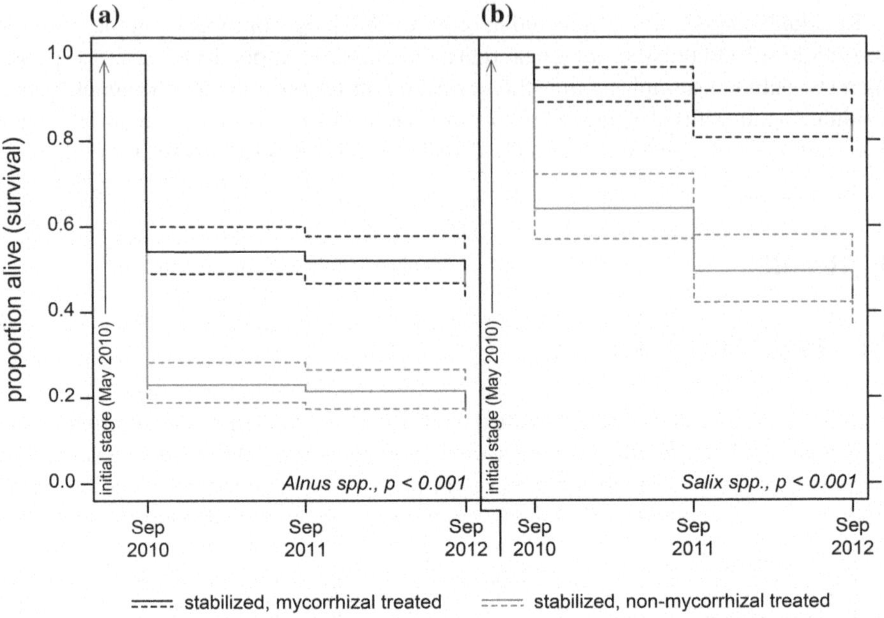

Fig. 3 The black (mycorrhized) and grey (control) solid lines show significant differences (p <0.001) for alder (**a**, n = 312 per site) and willow (**b**, n = 156 per site). Black and grey dashed lines illustrate the 95% confidence interval [51]

Fig. 4 Survival rates of *Salix herbacea* cuttings (n = 36 for each treatment) on a graded alpine ski slope (2535 m asl) depending on mycorrhizal inoculum type and the number of vegetation periods [40]

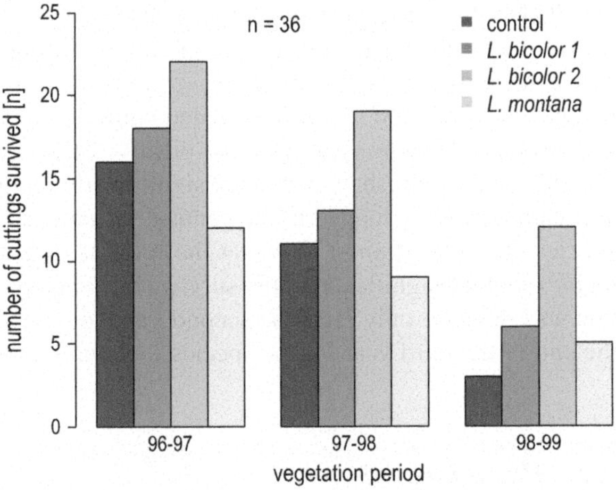

Differences between mean value and corresponding standard deviation (sd) and the robust median and mad (median absolute value) of shoot dry weight per sample, respectively, are negligible in all treatments. This suggests a lack of severe outliers

Fig. 5 Above ground biomass of *Betula pendula* with the commercial inoculum (left) and of a non-treated control (right) after 20 weeks of growing in the greenhouse

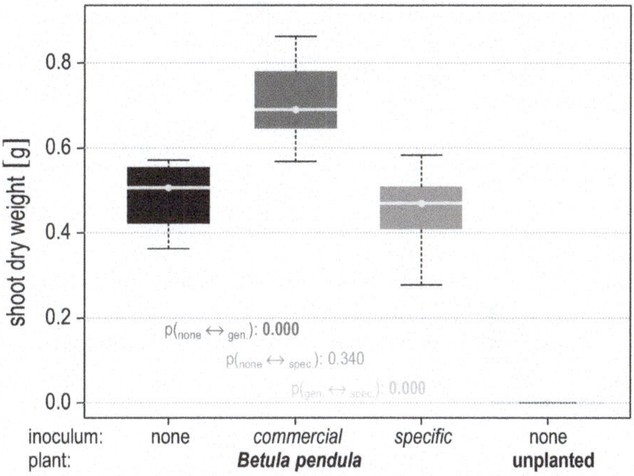

Fig. 6 Above ground biomass (dry weight) of *Betula pendula* plantlets (n = 12, per treatment) after the 20 weeks' growth period in the greenhouse measured as the root length per soil volume [cm cm^{-3}] for non-inoculated control (black), inoculated with the commercial (dark grey) and specific inoculum (grey) and the untreated soil (light grey)

and a good match to normal distribution assumptions as far as mean and sd are concerned. The corresponding values are given in Table 1.

Similar to the results of the above ground biomass, roots of the plants with the commercial inoculum grew significantly better and produced a much stronger branched root network compared to the control plants and those treated with the specific inoculum (Fig. 7).

Location and dispersion parameters for root length per soil volume per sample of the normal distribution (mean, sd) and their robust equivalents (median, mad) do not obviously differ for the control and specific inoculum. However, a considerable difference results for the treatment with the commercial inoculum, particularly, as

Table 1 Mean, standard deviation (sd), median, and median absolute value (mad) of shoot dry weight [g] per sample (cf. Fig. 6) in terms of soil planted with untreated as well as differently inoculated *Betula pendula* plantlets (n = 12, per treatment)

Shoot dry weight (g)	Mean	sd	Median	Mad
Betula untreated	0.489	0.077	0.507	0.089
Betula with commercial inoculum	0.703	0.095	0.689	0.107
Betula with specific inoculum	0.455	0.088	0.469	0.058

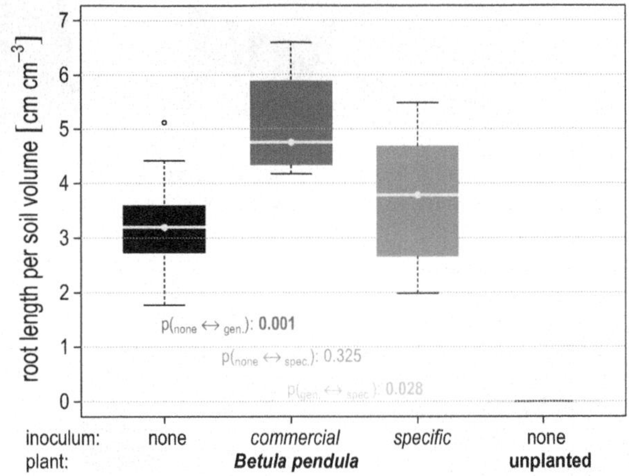

Fig. 7 The rooting performance of *Betula pendula* plantlets (n = 12, per treatment) after the 20 weeks' growth period in the greenhouse measured as the root length per soil volume [cm cm^{-3}] for non-inoculated control (black), inoculated with the commercial (dark grey) and specific inoculum (grey) and the untreated soil (light grey)

Table 2 Mean, standard deviation (sd), median, and median absolute value (mad) of root length density [cm cm^{-3}] per sample (cf. Fig. 7) in terms of soil planted with untreated as well as differently inoculated *Betula pendula* plantlets (n = 12, per treatment)

Root length density (cm $cm^{-3)}$	Mean	sd	Median	Mad
Betula untreated	3.205	1.005	3.200	0.890
Betula with commercial inoculum	5.375	1.506	4.750	0.815
Betula with specific inoculum	3.691	1.195	3.785	1.342

far as dispersion is concerned, with the standard deviation nearly double that of the robust mad (Table 2).

3.3 Soil Aggregate Stability

The soil aggregation potential of *Betula pendula* was significantly higher for both treatments with mycorrhiza compared to the control plants. Furthermore, the commercial inoculum performed again significantly better than the specific one. All planted treatments performed significantly better than bare soil (Fig. 8).

The comparison of mean and sd on the one hand with median and mad on the other hand of the soil aggregate stability revealed no noticeable differences within the different treatments, except for the dispersion parameters of the treatment commercial with the standard deviation nearly double that of the robust mad (Table 3).

Robust simple linear regression analysis for soil aggregate stability in dependence of root length per soil volume yielded significant positive correlation for the treatment with the specific inoculum (*p*-value: 0.025) but not for the non-inoculated plants (*p*-value: 0.191) and the treatment with the commercial inoculum (*p*-value: 0.836). Taking all treatments into account the model, again, demonstrates the significant

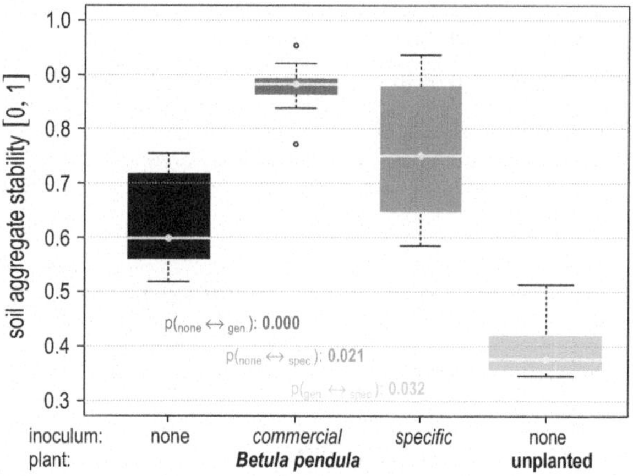

Fig. 8 Soil aggregation potential of *Betula pendula* plantlets (n = 12, per treatment) after the 20 weeks' growth period in the greenhouse measured as the root length per soil volume [cm cm^{-3}] for non-inoculated control (black), inoculated with the commercial (dark grey) and specific inoculum (grey) and the untreated soil (light grey)

Table 3 Mean, standard deviation (sd), median, and median absolute value (mad) of soil aggregate stability [0, 1] per sample (cf. Fig. 8) in terms of bare soil and soil planted with untreated as well as differently inoculated *Betula pendula* plantlets (n = 12, per treatment)

Soil aggregate stability [0, 1]	Mean	sd	Median	Mad
Control	0.369	0.055	0.376	0.042
Betula untreated	0.630	0.091	0.599	0.107
Betula with commercial inoculum	0.877	0.046	0.882	0.025
Betula with specific inoculum	0.761	0.124	0.751	0.174

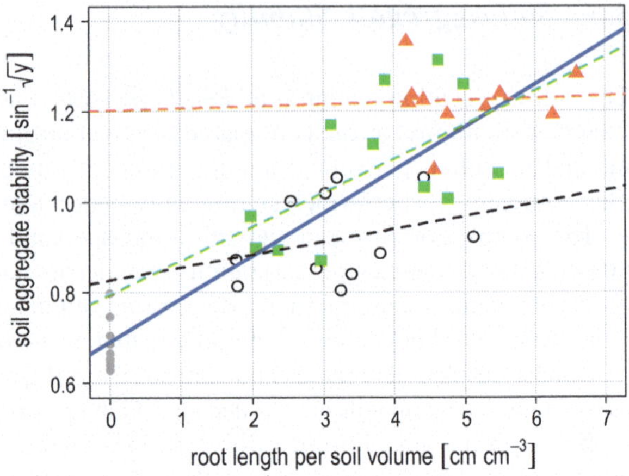

Fig. 9 Robust simple linear regression models $\sin^{-1}$(aggregate stability)$^{1/2}$ ~ root length for non-inoculated (dark grey, dashed), commercial (orange, dashed) and specific inoculum (green, dashed) as well as for the combination of all treatments of *Betula pendula* (blue, solid) including unplanted control (light grey points)

Table 4 Summary of the linear regression model $\sin^{-1}$(aggregate stability)$^{1/2}$ ~ root length including all treatments and control, showing significant difference in intercepts and slope

Coefficients	Estimate	Std. error	t value	Pr(>|t|)	
Intercept	0.6902	0.0172	40.13	$<2.0 \times 10^{-16}$	***
Root length	0.0950	0.0070	13.60	$<2.0 \times 10^{-16}$	***

Robust residual standard error: 0.1196
Multiple R-squared: 0.7324 Adjusted R-squared: 0.7259
Significance codes: ***<0.001; 0.001≤ **<0.01; 0.01≤ *<0.05

positive influence of roots of *Betula pendula* on soil aggregate stability (Fig. 9; Table 4).

3.4 Direct Shear Tests

Based on limit equilibrium analysis, the direct shear tests with planted (*Alnus incana*, *Poa pratensis*, *Trifolium pratense*) and unplanted large-scale specimens (50 cm × 50 cm × 30 cm) supported clearly the positive stabilisation effect of vegetation (Fig. 10). An average rooting ($\bar{c}_R$), yielding a reinforcement effect (cohesion) of about 6.5 kPa, enables slope stability up to 5° steeper (35°) than can be expected from the angle of internal friction Φ' of the bare soil material (30°), with the failure plane at a soil depth of 1.2 m [16]. A root reinforcement of ≈9 kPa, could theoretically increase slope stability up to 15°. These calculations are conducted for an infinite slope with seepage (parallel to the slope), based on limit equilibrium (Fig. 10).

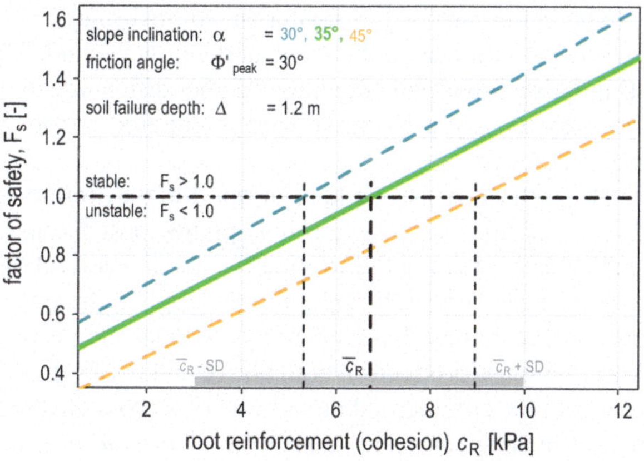

Fig. 10 Factor of safety calculation for slopes with an angle equal to (blue, dashed), 5° (green, solid), and 15° (gold, dashed) steeper than the friction angle Φ′ (30°) for a soil failure depth of 1.2 m with range ($\bar{c}_R \pm$ SD) of root reinforcement (cohesion). Calculations for an infinite slope with seepage (parallel to slope), based on limit equilibrium

4 Discussion

Depending on the magnitude of the soil degradation processes, mycorrhizal fungi are severely affected, yielding a lack of the mycorrhizal infection potential [52]. On harshly degraded soil, the natural recovery potential of the symbiotic fungi is considerably obstructed and, particularly, depends on the adjacent regions that are still intact. However, in this neighbourhood the species composition does not necessarily fit the requirements of the site to be re-colonised. In most cases the incompatibility is due to the difference in plant species found in the intact proximity and those applied for re-colonisation. As not every mycorrhizal fungus forms mycorrhiza with every plant, such discrepancy entails serious consequences in view of mycorrhization. A further impediment are successional processes in mycorrhizal communities in the way that perennial plants do have other fungal partners in their juvenile, prime, and senescent living phase [41, 53].

Recent research and experience suggest that the introduction of indigenous plant species together with a managed community of mycorrhizal fungi is an excellent approach in order to overcome the initial deficiencies in respect of a functional seedbed, stimulating and promoting autogenic recovery of degraded ecosystems [54–57]. This concept substantially increases sustainable revegetation success by creating synergies between abiotic and biotic processes and is advantageous not merely under extreme and limiting environmental conditions [58, 59].

The beneficial effect of applying mycorrhizal fungi on plant survival within the scope of the eco-engineering measures in the steep landslide area of the subalpine forest belt [20, 21] as well as on the alpine ski slope in Grisons [40] is in line with results of numerous other studies. Investigations conducted in different fields of applied mycorrhiza, e.g. plant nurseries, weaning of micropropagated plants, agriculture as well as re-colonisation of bare ground, consistently showed the positive effect of mycorrhizal fungi on survival of the corresponding host plants, too [58, 60–64].

The better survival of mycorrhized host plants is closely coupled to their better growth performance both above and below ground. The main reason for this growth stimulating function is the much higher efficiency of water and nutrient supply provided by the mycorrhizal network compared to root-hairs as exemplarily shown in Fig. 2 [28, 29].

However, it has to be kept in mind that different mycorrhizal fungi provide different functions for different host plants in different environments at different time steps [65] explaining the varying results on survival as well as root and shoot growth in dependence of the different inoculi applied. Inoculation with only *Hebeloma crustuliniforme*, although a known mycorrhizal fungus of birch, resulted in significantly lower biomass production both above and below ground as compared to the inoculation with the commercial inoculum. This product consists of 8 ecto- and 3 arbuscular mycorrhizal species and, apart from *H. crustuliniforme*, contains four other species, i.e. *Amanita muscaria* (L.) Lam., *Laccaria laccata* (Scop.) Cook., *Paxillus involutus* (Batsch) Fr., and *Thelephora terrestris*, Ehrh., known to naturally form mycorrhiza with birch [27, 66, 67]. The four additional fungal partners of birch most likely claimed responsibility for the significantly better growth performance and, consequently, the resulting increase in soil aggregate stability (Figs. 6, 7 and 8). It may, therefore, be speculated that *H. crustuliniforme* provides other beneficial functions for birch than particularly better and faster growth—e.g. protection against pathogens or toxic substances [26, 27]. The three arbuscular species of *Glomus* Tul. & C. Tul. included in the commercial inoculum are not of relevance—birch is exclusively colonised by ectomycorrhizal fungi [68].

Nevertheless, most mycorrhizal fungi known to naturally establish partnerships with plants used for stabilisation measures often yield an increase in survival as well as root growth and above ground biomass. However, now and then a more or less distinct time lag is observed until positive effects of mycorrhiza become obvious and significant as compared to untreated control plants. Furthermore, during the initial phase of mycorrhiza formation even a transient worse performance is possible [26].

Plant growth performance and, thus, mycorrhiza has further essential implications in respect of soil and slope stability as a key driver of both interception and evapo-transpiration. It has been scientifically proved under various conditions that above ground biomass in terms of leaf area index is positively correlated with interception of rainfall in forests as well as grassland [10, 13, 69, 70]. With regard to evapo-transpiration the vapour pressure deficit is a most influencing factor and it was demonstrated that it is reflected by aerodynamic resistance, representing surface roughness, the most influential feature of forest canopies [71]. According to the Penman-Montheith equation [72] it follows that evaporation decreases with increasing aerodynamic resistance which, on its part, decreases with increasing leaf area index [12, 73].

Consequently, an increase in above and below ground biomass increases both interception and evapo-transpiration and implies lower soil saturation levels. Thus, with regard to a heavy rainfall event, the soil has a higher water storage capacity and the time span until complete saturation is considerably extended.

In addition to the biomass depending effects of interception and evapo-transpiration, the resistance against pore water pressure is another decisive process in respect of soil and slope stability. It has been nicely shown that vegetation can reduce pore water pressure in soil by root water uptake resulting in higher shear strength but lower soil water permeability [74]. This soil stabilising effect is, again and obvious, positively correlated to plant growth in general and to root morphology and architecture in particular.

In a study on pore water pressure development during soil aggregate stability tests [75] it was found that the resistance against disintegration of soil aggregates by increasing pore water pressure—measured as the delay until passing from negative to positive pore water pressure—holds longer in samples with higher root length density and finer root networks. Consequently, soil aggregate stability was significantly higher after a plant growth period of about 20 weeks in the samples of the grass *Poa pratensis* in contrast to those of the tree *Alnus incana* for which considerably lower root length density and a much higher portion of middle to coarse roots was observed. Thus, it can be speculated that particularly the portion of fine roots contributes to the retarded increase in pore water pressure at least in a very early phase of plant and soil development. Ng et al. [74] found that exponential root architecture induces the highest negative pore water pressure in the soil followed by triangular, uniform, and parabolic architecture. Roots of *P. pratensis* follow more the exponential architecture whereas those of *A. incana* are more uniform like. This is further evidence for a stronger reducing effect of pore water pressure and an increase in soil aggregate stability by the grass species. Additionally, at equal treatment, *P. pratensis* had higher evapo-transpiration efficiency, indicated by a lower pore water pressure level at the start of the experiments. Therefore, it is conceivable that pore water pressure is associated to soil aggregate stability the same way as root length density, since the latter two are positively correlated as seen in Fig. 9 and demonstrated in numerous other studies, e.g. [20, 25, 38, 43, 76]. The close interrelationship between soil aggregate stability and pore water pressure is anything else than surprising as an increase in the latter provokes disruption and slaking of the former [77].

The direct and indirect effects of roots as well as their associated mycorrhizal fungi contribute to the stability of soil aggregates and, therefore, to the strength of the soil matrix and resistance against breaking [20, 25, 38, 43]. Apart from increasing root length density and chemical cementation by polysaccharides, the mycorrhizal fungi further contribute to soil and slope stability inducing morphogenetic changes in root architecture and increasing their 3D complexity (Fig. 2). It is assumed that a positive correlation exists between root length density and ramification with lateral roots [78]. Therefore, with the mycorrhization of plants, a pronounced increase in the complexity of the roots' 3D-structure is expected. Up to a certain degree, the higher soil aggregate stability provided by birch mycorrhized with the commercial inoculum is most likely due to a gain in 3D complexity, too. A corresponding increase in shear strength can be assumed as certain and was confirmed with triaxial compression tests. Investigations on sand reinforced with artificial 3D inclusions revealed a positive correlation between shear strength and the complexity of the 3D structures added [79]. This is similar to the results of triaxial compression tests of planted moraine, where those specimens with the highest root length density and, correspondingly, most-branched root systems mobilised the maximum peak shear force mirrored by the angle of internal friction Φ' [19].

A resilient soil matrix consisting of stable micro- and macro-aggregates assembled by fungal mycelia and roots is not only a prerequisite for slope stability but provides a significantly higher retention capacity for water and nutrients. In turn, mycorrhized compared to non-mycorrhized plants have much greater access to this enriched nutrient pool yielding better growth performance and survival rates resulting in the development and establishment of a protective vegetation cover. This positive feedback-loop is the central key and essential requirement in respect of successful eco-engineering measures aimed at sustainable slope stabilisation and protection against shallow landslides.

From the soil mechanical perspective, calculations for an infinite slope with seepage (parallel to the slope), based on limit equilibrium applying the data of direct shear tests with large-scale specimens (50 cm × 50 cm × 30 cm) generally support the aforementioned findings. In respect of plant functions on slope stability, particularly the positive correlation between roots and additional strength of soil was demonstrated as shown in Fig. 10 [16]. A difference of 5° between the slope angle α and the angle of internal friction Φ', reflected by an increase in peak shear force, results from a mean root reinforcement (cohesion) of about 6.5 kPa achieved by planting the moraine with the tree species *Alnus incana*, the grass *Poa pratensis* and the legume *Trifolium pratense*. On slopes covered with well-maintained protection forests such values can be assumed [80] at a depth of 1.2 m for the potential failure plane, representing the mean of more than 200 shallow landslides triggered in forests [81]. The 5° increase in the angle of internal friction Φ' is further confirmed by triaxial compression tests of the same soil material, planted with *A. incana*, and in line with the soil aggregate stability approach [19, 43].

It can be speculated that protection forests meeting the optimal requirements of NaiS, the guidelines for "Sustainability in Protection Forests" [82], keep slopes stable at even greater differences between the angle of internal friction Φ' of bare soil and the effective slope inclination α (Fig. 10). This assumption is based on the soil mechanical experiments with root-permeated soil [16] and investigations on how forest structure affects root reinforcement and susceptibility to shallow landslides [80].

5 Conclusions

The hostile conditions for vegetation development on severely degraded slopes and the beneficial effects of mycorrhizal fungi elucidate that the application of plants together with their relevant fungal partners is highly recommendable. Mycorrhizal fungi considerably and positively contribute to plant growth, survival, soil stability and corresponding slope stability functions of plants, e.g. interception and evapotranspiration. However, the selection of the fungal species needs to experience careful examination. Sound information on their ecology and sociology as well as on the potential for aggregate formation is required depending on the plant species used as initial step in the re-colonisation process as well as in view of the climax association in mind.

Considering all these aspects, the application of mycorrhizal inoculi in eco-engineering will be undoubtedly an important step towards more sustainability and absolutely merits to be integrated into engineering concepts and risk management.

Acknowledgements We thank the Swiss National Research Programme 68 (grant number 406840_143122, SOSTANAH), the EU project REVENUES (FP7-PEOPLE-2012-IAPP; grant number 324466) as well as the BAFU (no 09.0027.PJ/I211-3446) and the "Wolfermann Nägeli Stiftung Zürich" for supporting and funding our research.

Appendix

Residual analysis of the model: $\sin^{-1}(\text{aggregate stability})^{½} \sim$ root length (Fig. 9; Table 4).

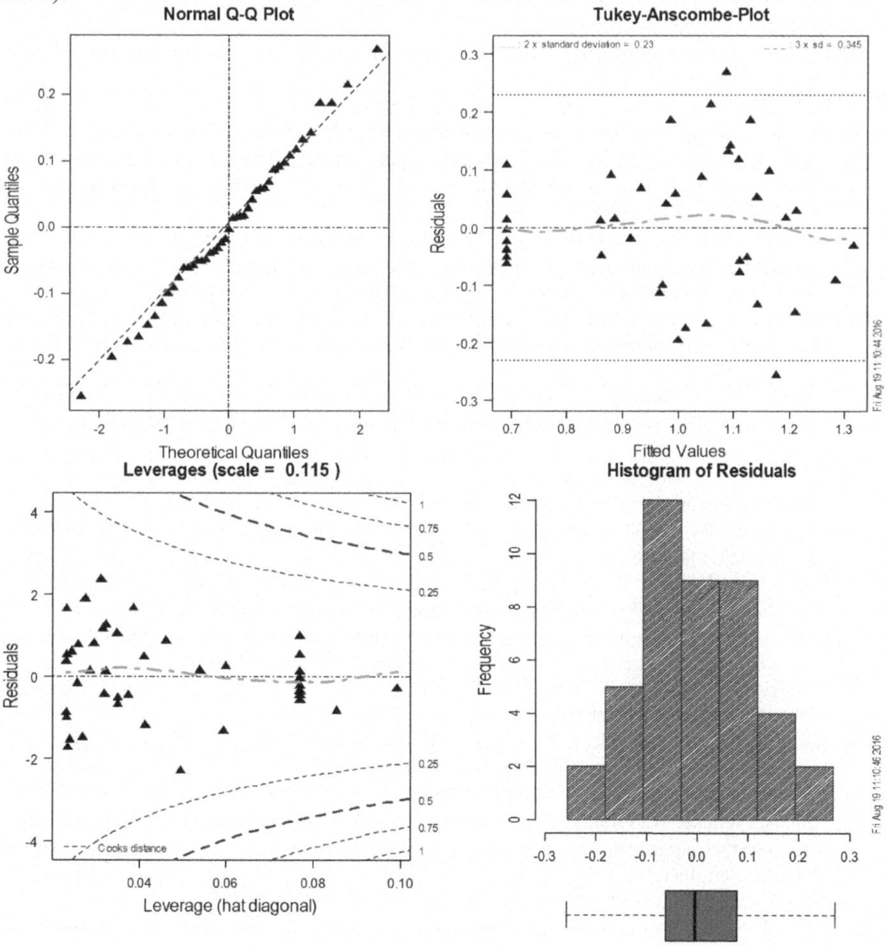

References

1. Beniston, M., Stoffel, M., Hill, M.: Impacts of climatic change on water and natural hazards in the Alps: can current water governance cope with future challenges? Examples from the European "ACQWA" project. Environ. Sci. Policy **14**, 734–743 (2011)
2. Bezzola, G.R., Hegg, C.: Ereignisanalyse Hochwasser 2005. Teil 1 – Prozesse, Schäden und erste Einordnung. Bern, Bundesamt für Umwelt BAFU, 215 S. Eidgenössische Forschungsanstalt WSL, Birmensdorf (2007)
3. Chloros, N.: Enzyklopädie der Forstwirtschaft. Griechisch, Athen (1891)
4. Kontos, P.: Griechische Forstgeschichte. Griechisch, Athen (1929)
5. Goulas, K., Graf, F.: Lebendverbau in Wildbacheinzugsgebieten Griechenlands. Ingenieurbiologie **4**, 19–25 (2003)
6. WAG: Bundesgesetz über den Wald (Waldgesetz, WaG), vom 4. SR 921.0, Oktober 1991
7. Hörman, G., Branding, A., Clemen, T., Herbst, M., Hinrichs, A.: Calculation and simulation of wind controlled canopy interception of beech forest in Northern Germany. Agric. For. Meteorol. **79**, 131–148 (1996)
8. Loshali, D.C., Singh, R.P.: Partitioning of rainfall by three Central Himalayan forests. For. Ecol. Manage. **53**, 99–105 (1992)
9. Liu, S.: A new model for the prediction of rainfall interception in forest canopies. Ecol. Model. **99**, 151–159 (1997)
10. Park, H.: Physical characteristics of heat and water exchange processes between vegetation and the atmosphere in a deciduous broad-leaved forest. Nagoya University (2000)
11. Price, A.G., Carlyle-Moses, D.E.: Measurement and modelling of growing-season canopy water fluxes in a mature mixed deciduous forest stand, southern Ontario, Canada. Agric. For. Meteorol. **119**, 69–85 (2003)
12. Deguchi, A., Hattori, S., Park, H.T.: The influence of seasonal changes in canopy structure on interception loss: application of the revised Gash model. J. Hydrol. **318**, 80–102 (2006)
13. Aston, A.R.: Rainfall interception by eight small trees. J. Hydrol. **42**, 383–396 (1979)
14. Querejeta, J.I., Egerton-Warburtona, L.M., Allen, M.F.: Hydraulic lift may buffer rhizosphere hyphae against the negative effects of severe soil drying in a California Oak savanna. Soil Biol. Biochem. **39**, 409–417 (2008)
15. Bauerle, T.L., Richards, J.H., Smart, D.R., Eissenstat, D.M.: Importance of internal hydraulic redistribution for prolonging the lifespan of roots in dry soil. Plant Cell Environ. **31**, 177–186 (2008)
16. Yildiz, A., Askarinejad, A., Graf, F., Rickli, C., Springman, S.M.: Effects of roots and mycorrhizal fungi on the stability of slopes. In: Proceedings of the XVI ECSMGE Geotechnical Engineering for Infrastructure and Development, Edinburgh, pp. 1693–1698 (2015). doi: https://doi.org/10.1680/ecsmge.60678
17. Frei, M.: Validation of a new approach to determine vegetation effects on superficial soil movements. ETH (2009). https://doi.org/10.3929/ethz-a-005954329
18. Böll, A., Graf, F.: Nachweis von Vegetationswirkungen bei oberflächennahen Bodenbewegungen – Grundlagen eines neuen Ansatzes. Schweiz. Z. Forstwes. **152**, 1–11 (2001)
19. Graf, F., Frei, M., Böll, A.: Effects of vegetation on the angle of internal friction of a moraine. FOSNOLA **82**, 61–78 (2009)
20. Bast, A., Wilcke, W., Graf, F., Lüscher, P., Gärtner, H.: The use of mycorrhiza for eco-engineering measures in steep alpine environments: effects on soil aggregate formation and fine-root development. Earth Surf. Proc. Land. **13**, 1753–1763 (2014)
21. Bast, A., Wilcke, W., Graf, F., Lüscher, P., Gärtner, H.: Does mycorrhizal inoculation improve plant survival, aggregate stability, and fine root development on a coarse-grained soil in an alpine eco-engineering field experiment? J. Geophys. Res. Biogeosci. (2016). https://doi.org/10.1002/2016JG003422
22. Wu, T.H.: Effect of vegetation on slope stability. Transp. Res. Rep. **965**, 37–46 (1984)

23. Pollen, N., Simon, A.: Estimating the mechanical effects of riparian vegetation on stream bank stability using a fibre bundle model. Water Resour. Res. **41**, W07025 (2005). https://doi.org/10.1029/2004wr003801
24. Schwarz, M., Cohen, D., Or, D.: Soil-root mechanical interactions during pullout and failure of root bundles. J. Geophys. Res. **115**, F04035 (2010). https://doi.org/10.1029/2009JF001603
25. Bast, A., Wilcke, W., Graf, F., Lüscher, P., Gärtner, H.: A simplified and rapid technique to determine an aggregate stability coefficient in coarse grained soils. CATENA **127**, 170–176 (2015)
26. Smith, S.E., Read, D.J.: Mycorrhizal Symbiosis, p. 787. Academic Press, London (2008)
27. Cairnay, J.W.G., Chambers, S.M.: Ectomycorrhizal Fungi. Key Genera in Profile, 369 pp. Springer, Berlin, (1999)
28. Read, D.J., Boyd, R.: Water relations of mycorrhizal fungi and their host plants. In: Ayres, P.G., Boddy, L. (eds.) Water, Fungi and Plants, pp. 287–303. Cambridge University Press, Cambridge (1986)
29. Jansen, A.E.: Importance of ectomycorrhiza for forest ecosystems. In: Teller, A., Mathy, P., Jeffers, J.N.R. (eds.) Response of Forest Ecosystems to Environmental Changes, pp. 456–461. Elsevier, London and New York (1992)
30. Budi, S.W., van Tuinen, D., Martinotti, G., Gianinazzi, S.: Isolation from the *Sorghum bicolor* mycorrhizasphere of a bacteria compatible with arbuscular mycorrhiza development and antagonistic towards soilborne fungal pathogens. Appl. Environ. Microbiol. **65**, 5148–5150 (1999)
31. Filion, M., St-Arnaud, M., Fortin, J.A.: Direct interaction between the arbuscular mycorrhizal fungus *Glomus intraradices* and different rhizosphere micro-organisms. New Phytol. **141**, 525–533 (1999)
32. Bezzate, S., Aymerich, S., Chambert, R., Czarnes, S., Berger, O., Heulin, T.: Disruption of the *Paenibacillus polymyxa* levansucrase gene impairs its ability to aggregate soil in the wheat rhizosphere. Environ. Micorbiol. **2**, 333–342 (2000)
33. Hildebrandt, U., Janetta, K., Bothe, H.: Towards growth of arbuscular mycorrhizal fungi independent of a plant host. Appl. Environ. Microbiol. **68**, 1919–1924 (2002)
34. Mansfeld-Giese, K., Larsen, J., Bodker, L.: Bacterial populations associated with mycelium of the arbuscular mycorrhizal fungus *Glomus intraradices*. FEMS Micorbiol. Ecol. **41**, 133–140 (2002)
35. Caesar-Ton That, T.C., Shelver, W.L., Thorn, R.G., Cochran, V.L.: Generation of antibodies for soil aggregating basidiomycete detection as an early indicator of trends in soil quality. Appl. Soil. Ecol. **18**, 99–116 (2001)
36. Tagu, D., de Bellis, R., Balestrini, R., de Vries, O.M.H., Piccoli, G., Stocchi, V., Bonfante, P., Martin, F.: Immuno-localization of hydrophobin HYDPt-1 from the ectomycorrhizal basidiomycete *Pisolithus tinctorius* during colonization of *Eucalyptus globulus* roots. New Phytol. **149**, 127–135 (2001)
37. Mankel, A., Krause, K., Kothe, E.: Identification of hydrophobin gene that is developmentally regulated in the ectomycorrhizal fungus *Tricholoma terreum*. Apppl. Environ. Microbiol. **68**, 1408–1413 (2002)
38. Rillig, M.C., Mummey, D.L.: Mycorrhizas and soil structure. New Phytol. **171**, 41–53 (2006)
39. INOQ GmbH: Solkau 2, 29465 Schnega, Deutschland. Mykorrhiza Produkte – Forst (2016). http://inoq.de/produkte-service/mykorrhiza-produkte/inoq-forst/
40. Graf, F.: Anforderungen an den Boden bei Renaturierungen oberhalb der Waldgrenze. Ingenieurbiologie **4**, 12–18 (1998)
41. Graf, F.: Ecology and sociology of macromycetes in snow-beds with *Salix herbacea* L. in the alpine Valley of Radönt (Grisons, Switzerland). Diss. Bot. **235**, 1–242 (1994)
42. WinRhizo®: Régent Instruments Inc., 4040 rue Blain, Quebec, Qc G2B 5C3, Canada (2000). http://www.regent.qc.ca
43. Graf, F., Frei, M.: Soil aggregate stability related to soil density, root length, and mycorrhiza using site-specific *Alnus incana* and *Melanogaster variegatus s.l.* Ecol. Eng. **57**, 314–323 (2013)

44. SN 670 010: Geotechnische Erkundung – Geotechnische Kenngrössen. Schweizerischer Verband der Strassen-und Verkehrsfachleute VSS, Schweizer Norm, 19 S (2011)
45. Germann, P., Helbling, A., Vadilonga, T.: Rivulet approach to rates of preferential infiltration. Vadose Zone J. **6**, 207–220 (2007)
46. R Development Core Team: A language and environment for statistical computing. R Foundation for Statistical Computing, Vienna, Austria. ISBN 3-900051-07-0 (2016). http://www.R-project.org
47. Kaplan, E.L., Meier, P.: Nonparametric estimation from incomplete observations. J. Am. Stat. Assoc. **53**, 457–481 (1958). https://doi.org/10.1080/01621459.1958.10501452
48. Tableman, M., Kim, J.S.: Survival Analysis Using S: Analysis of Time-to-Event Data, 280 pp. Chapman & Hall/CRC Texts in Statistical Science, CRC Press, Boca Raton, Florida (2003)
49. Harrington, D.P., Fleming, T.R.: A class of rank test procedures for censored survival data. Biometrika **69**, 553–566 (1982). https://doi.org/10.1093/biomet/69.3.553)
50. Tukey, J.W.: On the comparative anatomy of transformations. Ann. Math. Stat. **28**, 602–632 (1987)
51. Bast, A.: Mycorrhizal inoculation as a promoter for sustainable eco-engineering measures in steep alpine environments? Diss. Univ. Bern, 149 S (2014)
52. Amaranthus, M.P., Trappe, J.M.: Effects of erosion on ecto- and VA-mycorrhizal inoculum potential of soil following forest fire in southwest Oregon. Plant Soil **150**, 41–49 (1993)
53. Last, F.T., Mason, P.A., Wilson, J., Deacon, J.W.: Fine roots and sheathing mycorrhizas: their formation, function and dynamics. Plant Soil **71**, 9–21 (1983)
54. Requena, N., Perez-Solis, E., Azcon-Aguilar, C., Jeffries, P., Barea, J.M.: Management of indigenous plant-microbe symbioses aids restoration of desertified ecosystems. Appl. Environ. Microbiol. **67**, 495–498 (1983)
55. Caravaca, F., Barea, J.M., Palenzuela, J., Figueroa, D., Alguacil, M.M., Roldan, A.: Establishment of shrub species in a degraded semiarid site after inoculation with native or allochthonous arbuscular mycorrhizal fungi. Appl. Soil. Ecol. **22**, 103–111 (2003)
56. King, E.G., Hobbs, R.J.: Identifying linkages among conceptual models of ecosystem degradation and restoration: towards an integrative framework. Restor. Ecol. **14**, 369–378 (2006)
57. Chaudhary, V.B., Bowker, M.A., O'Dell, T.E., Grace, J.B., Redman, A.E., Rillig, M.C., Johnson, N.C.: Untangle the biological contributions to soil stability in semiarid shrublands. Ecol. Appl. **19**, 110–122 (2009)
58. Jeffries, P., Gianinazzi, S., Perotto, S., Turnau, K., Barea, J.M.: The contribution of arbuscular mycorrhizal fungi in sustainable maintenance of plant health and soil fertility. Biol. Fertil. Soils **37**, 1–16 (2003)
59. Byers, J.E., Cuddington, K., Jones, C.G., Talley, T.S., Hastings, A., Lambrinos, J.G., Crooks, J.A., Wilson, W.G.: Using ecosystem engineers to restore ecological systems. Trends Ecol. Evol. **21**, 493–500 (2006)
60. Grime, J.P., Mackey, J.M., Hillier, S.M., Read, D.J.: Floristic diversity in a model system using experimental microcosms. Nature **328**, 420–422 (1987)
61. Jakobsen, I., Joner, E.J., Larsen, J.: Hyphal phosphorus transport, a keystone to mycorrhizal enhancement of plant growth. In: Gianinazzi, S., Schüepp, H. (eds.) Impact of Mycorrhizal Colonisation on Root Architecture, Root Longevity and the Formation of Growth Regulators, pp. 133–146. Birkhäuser Verlag, Basel (1994)
62. Berta, G., Trotta, A., Fusconi, A., Hooker, J.E., Munro, M., Atkinson, D., Giovannetti, M., Morini, S., Fortuna, P., Tisserant, B., Gianinazzi-Pearson, V., Gianinazzi, S.: Arbuscular mycorrhizal induced changes to plant growth and root system morphology in *Prunus cerasifera*. Tree Physiol. **15**(5), 281–293 (1995). https://doi.org/10.1093/treephys/15.5.281
63. Martins, A., Barroso, J., Pais, M.S.: Effect of ectomycorrhizal fungi on survival and growth of micropropagated plants and seedlings of *Castanea sativa* Mill. Myorrhiza **6**, 265–270 (1996)
64. Nara, K., Hogetsu, T.: Ectomycorrhizal fungi on established shrubs facilitate subsequent seedling establishment of successional plant species. Ecology **85**, 1700–1707 (2004)
65. Perry, D.A., Molina, R., Amaranthus, M.P.: Mycorrhizae, mycorrhizospheres, and reforestation: current knowledge and research needs. Can. J. For. Res. **17**, 929–940 (1987)

66. Fox, F.: Groupings of ectomycorrhizal fungi of birch and pine, based on establishment of mycorrhiza on seedlings from spores in unsterile soil. Trans. Br. Mycol. Soc. **87**, 371–380 (1986)
67. Gibson, F., Deacon, J.W.: Experimental study of establishment of ectomycorrhizas in different regions of birch root systems. Trans. Br. Mycol Soc. **91**, 239–251 (1988)
68. Harley, J.L., Harley, E.L.: A check-list of mycorrhizae in the British Flora. New Phytol. **105**, 1–102 (1987)
69. Crockford, R.H., Richardson, D.P.: Partitioning of rainfall into throughfall, stem flow and interception: effect of forest type, ground cover and climate. Hydrol. Process. **14**, 2903–2920 (2000)
70. Gigante, V., Iacobellis, V., Manfreda, S., Milelli, P., Porthogese, I.: Influences of Leaf Area Index estimations on water balance modeling in a Mediterranean semi-arid basin. Nat. Hazards Earth Syst. Sci. **9**, 979–991 (2009)
71. Jarvis, P.G., Stewart, J.B.: Evaporation of water from plantation forest. In: Ford, E.D., Malcolm, D.C., Atterson, J. (eds.) The Ecology of Even-aged Forest Plantations. Institute of Terrestrial Ecology, Cambridge (1978)
72. Gash, J.H.C., Morton, A.J.: An application of the Rutter model to the estimation of the interception loss from Thetford Forest. J. Hydrol. **38**, 49–58 (1978)
73. Teklehaimanot, Z., Jarvis, P.G., Ledger, D.C.: Rainfall interception and boundary layer conductance in relation to tree spacing. J. Hydrol. **123**, 261–278 (1991)
74. Ng, C.W.W., Liu, H.W., Feng, S.: Analytical solutions for calculating pore-water pressure in an infinite unsaturated slope with different root architectures. Can. Geotech. J. **52**, 1981–1992 (2015)
75. Bader, A.: How plants and mycorrhizal fungi contribute to soil aggregate stability. Master thesis, 146 pp. ETH (2014). http://www.wsl.ch/fe/gebirgshydrologie/wildbaeche/ projekte/SOSTANH/Theses_EN/AnjaBader
76. Pohl, M., Graf, F., Butler, A., Rixen, C.: The relationship between plant species richness and soil aggregate stability can depend on disturbance. Plant Soil **355**, 87–102 (2012)
77. LeBissonnais, Y.: Aggregate stability and assessment of soil crustability and erodibility. I. Theory and methodology. Eur. J. Soil Sci. **47**, 425–437 (1996)
78. Berta, G., Fusconi, A., Trotta, A., Cannerini, S.: Morphogenetic modifications induced by the mycorrhizal fungus *Glomus* strain E_3 in the root system of *Allium porrum* L. New Phytol. **114**, 207–215 (1990)
79. Zhang, M.X., Javadi, A.A., Min, X.: Triaxial tests of sand reinforced with 3D inclusions. Geotext. Geomembr. **24**, 201–209 (2006)
80. Moos, C., Bebi, P., Graf, F., Mattli, J., Rickli, C., Schwarz, M.: How does forest structure affect root reinforcement and susceptibility to shallow landslides? Earth Surf. Process. Landforms **41**, 951–960 (2016)
81. Rickli, C. (ed.): Vegetationswirkungen und Rutschungen. Untersuchung zum Einfluss der Vegetation auf oberflächennahe Rutschprozesse anhand der Unwetterereignisse in Sachseln OW am 15. August 1997. Birmensdorf, Bern, Eidg. Forschungsanstalt WSL, Bundesamt für Umwelt, Wald und Landschaft. 97 S (2001)
82. Frehner, M., Wasser, B., Schwitter, R.: Nachhaltigkeit und Erfolgskontrolle im Schutzwald. Wegleitung für Pflegemassnahmen in Wäldern mit Schutzfunktion, Vollzug Umwelt. Bundesamt für Umwelt, Wald und Landschaft, Bern, 564 S (2005)

Shear Strength of Granular Soil Under Saturated and Unsaturated Conditions

Gregor Idinger and Wei Wu

Abstract The shear strength of soils depends on various factors such as the stress level, soil density, and degree of saturation. Capillary forces in unsaturated soils feature an increased strength against shear loading, and influence stress, deformation, and flow phenomena. Thus it is crucial to incorporate the bonding effect of unsaturated soils to numerical models addressing the behaviour of unsaturated soils. The presented experimental campaign of triaxial compression tests aims to investigate the effect of capillarity on the mechanical behaviour of granular soil. The experimental programme consists of consolidated-drained tests on saturated soils and equivalent constant suction tests on unsaturated soils with matric suction of 10 kPa. The results of the experimental campaign reflect the higher resistance of unsaturated soils against shear loading manifested by an additional apparent cohesion term and by suction-induced hardening.

1 Introduction

Soil water plays a key role for the stability of geotechnical structures. While positive pore water pressures have negative influence on the soil strength, negative pore pressures in partially saturated soils provide additional strength due to "bonding" effects of capillary forces. Classical theories of limit state soil mechanics implicitly assume that soils are either dry or saturated. However, soil near the ground surface is commonly in the unsaturated state in between these two extreme conditions. Thus,

G. Idinger (✉) · W. Wu
Institute of Geotechnical Engineering, University of Natural Resources and Life Sciences, Feistmantelstrasse 4, 1180 Vienna, Austria
e-mail: gregor.idinger@boku.ac.at

W. Wu
e-mail: wei.wu@boku.ac.at

W. Wu (ed.), *Recent Advances in Geotechnical Research*,
Springer Series in Geomechanics and Geoengineering,
https://doi.org/10.1007/978-3-319-89671-7_7

for many geotechnical applications the unsaturated condition of soils influences the performance of engineering structures.

Soils are heterogeneous geomaterials composed of the solid soil skeleton and the void space. Voids are filled with liquid and gaseous phases. Thereof, unsaturated soils are idealised by a three-phase model, with solids, water, and air. The soil skeleton consists of particles with solid density ρ_s, and its volumetric fraction is defined by the ratio of volume of void space per volume of solid fraction as the void ratio $e = V_v/V_s$. The void space occupied by the pore water V_w is indicated by the degree of saturation $S_r = V_w/V_v$. Water molecules of the air-water interface are attracted towards the liquid phase causing an unbalanced force. The generated surface tension T_s achieves equilibrium conditions along the contractile skin. This tensile force interacts with the soil particles and thereof controls the mechanical behaviour of the soil skeleton. In the capillary model the radius of curvature of the water-air interface is inversely proportional to the difference between air pressure u_a and water pressure u_w within the soil. The difference in pressure is called matric suction $s = u_a - u_w$. For multiphase porous media such as soils it is crucial to address those effects in constitutive models formulated for unsaturated soils.

The objective of the presented experimental campaign is to describe and quantify the coupled shear and volumetric behaviour of saturated and unsaturated specimen of granular soil.

2 Material and Methods

The experimental investigation is conducted on granular soil. The coupled shear and volumetric behaviour is determined experimentally with the double-cell triaxial apparatus on infinite small homogeneous soil specimens for both saturated and unsaturated conditions. The triaxial testing programme consists of consolidated-drained tests on saturated soils and equivalent constant suction tests on unsaturated soils.

2.1 Soil Material

The soil is classified as well-graded medium-fine sand with a minor fraction of silt (mfSa, si′). Figure 2 presents the grain size distribution curve. The mean diameter is $d_{50} = 0.23$ mm, the coefficient of uniformity $C_u = 13.6$ and the coefficient of curvature $C_c = 3.6$. The scanning-electron-microscope image of Fig. 1 visualizes the sub-angular shape of sand particles. The particle density of the medium-fine sand is determined with $\rho_s = 2.67\,\mathrm{g/cm^3}$, quartz sand is the main component. The minimum and maximum void ratio are defined with $e_{min} = 0.45$ and $e_{max} = 1.09$, respectively. Soil specimens are compacted to medium dense and very dense states, as listed in the triaxial testing programme of Table 1.

2.2 Double Cell Triaxial Apparatus

The concept of the double-cell triaxial apparatus introduced by Simon J. Wheeler in 1986 enables the measurement of volumetric strain of unsaturated soil specimen [6]. The axis translation technique is adopted to control matric suction inside the specimen by imposing a positive air pressure u_a and keeping the difference in global pressure $(u_a - u_w)$ unchanged.

The employed double-cell triaxial apparatus manufactured by *Wille Geotechnik* is illustrated in Fig. 3. The major principal stress σ_1 in axial direction is measured by the submerged load cell, the axial displacement is measured by the LVDT mounted on top of the triaxial cell. The pore air pressure u_a is injected through the porous stone on top of the specimen. The pore water pressure u_w is controlled through the lower ceramic disc with high air-entry-value at the bottom of the specimen, and is measured by the pressure transducer connected via the lower ceramic disc. The confining pressure σ_3 is provided and regulated by an automatic displacement-controlled hydraulic pressure system. The global volumetric strain ε_V is inferred from the ram displacement of the automatic pressure system for the inner cell to preserve constant confining pressure.

Fig. 1 Scanning electron microscope image of the granular soil

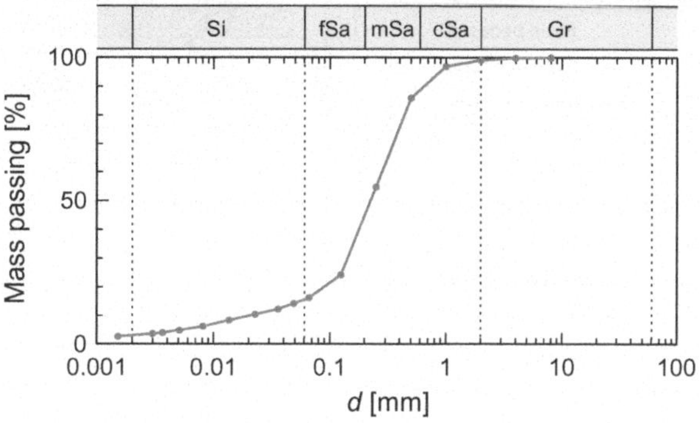

Fig. 2 Grain size distribution curve of the granular soil. Gr: 0.8%, Sa: 82.9%, Si: 13.4%, Cl: 2.8%

Table 1 Triaxial test programme with initial and failure conditions. Encoding of test identity: S - saturated, U - unsaturated; M - medium dense, D - very dense

Test	σ_3 (kPa)	e (–)	s (kPa)	w_0 (%)	w_f (%)	$S_{r,f}$ (%)	m (g)	p'_p (kPa)	q_p (kPa)	E (MPa)
SM	50	0.73	0.0	15.1	22.3*	81.2*	2958	73.2	93.1	10.0
	100	0.73	0.0	15.1	23.8*	86.6*	2964	170.1	220.1	11.5
	200	0.73	0.0	15.1	22.0*	81.3*	2969	343.5	438.7	14.6
UM	50	0.75	9.8	15.1	10.5	37.6	2931	79.2	110.9	14.5
	100	0.73	9.7	15.1	10.5	38.8	2968	169.8	230.5	16.0
	200	0.73	9.7	15.1	11.2	40.8	2961	342.1	450.6	18.2
SD	50	0.51	0.0	16.6	19.0	99.8	3426	121.9	218.4	30.7
	100	0.51	0.0	16.6	18.7	96.6	3448	233.3	404.2	57.3
	200	0.52	0.0	16.6	18.1	93.5	3418	431.6	699.4	68.0
UD	50	0.50	9.1	15.1	11.3	60.4	3409	118.7	224.1	30.0
	100	0.51	7.3	15.1	11.7	60.5	3385	233.8	415.8	49.0
	200	0.51	9.3	15.1	11.0	57.8	3396	437.6	730.5	69.6

*Drainage during unloading resulted in decrease of w_f and $S_{r,f}$

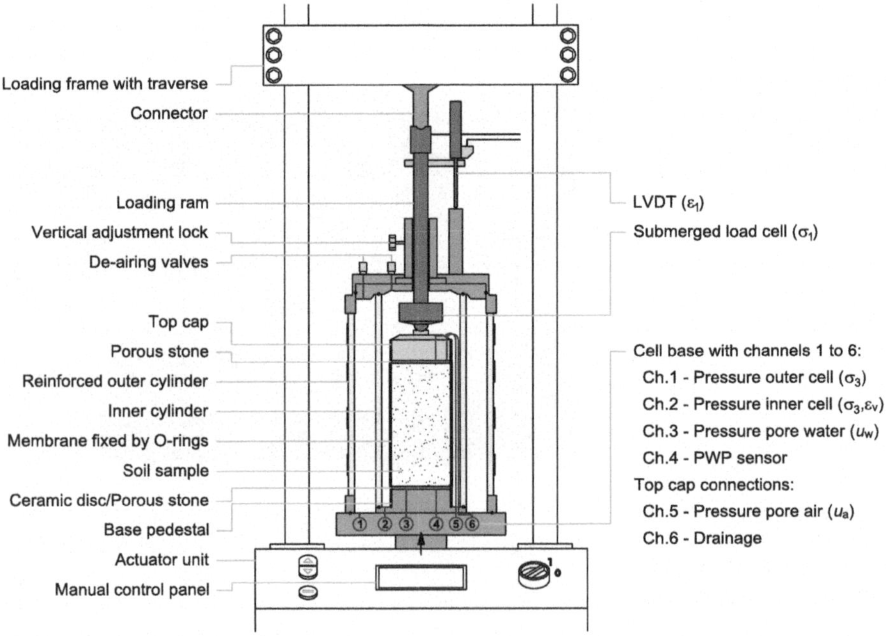

Fig. 3 Schematic drawing of the triaxial apparatus equipped with double cell configuration

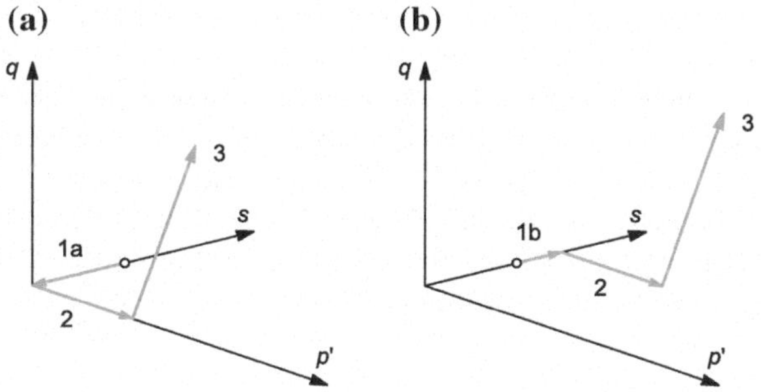

Fig. 4 Stress paths in the p'–q–s space for **a** saturated and **b** unsaturated triaxial compression tests, with 1a - initial saturation, 1b - setting of matric suction, 2 - consolidation, and 3 - drained axial compression

2.3 *Testing Procedure*

The triaxial testing programme consists of consolidated–drained tests for saturated specimen and equivalent constant suction tests for unsaturated specimen. The testing programme is summarized in Table 1. The stress paths are sketched in the p'–q–s space diagram of Fig. 4. Saturated soil specimen are initially saturated along path 1a with a back-pressure of 300 kPa. Unsaturated soil specimen are set to the desired matric suction along path 1b by applying the axis-translation technique. Soil specimens are of cylindrical shape with 100 mm in diameter and 200 mm in height. The soil is mixed at the optimum water content of $w_{opt} = 15.1\%$ and compacted by the moist-tamping technique in eight layers.

Four sets of triaxial tests are carried out: for medium dense $e = 0.74$ and very dense $e = 0.51$ compaction states under both saturated and unsaturated condition. The matric suction in the unsaturated tests is $s = 10$ kPa. This value corresponds to the intersection of the transition and residual zone of the SWCC [3]. Each set of tests consists of three individual tests isotropically consolidated to 50, 100, and 200 kPa. After reaching constant conditions, axial compression is applied at the strain rate of 0.02 mm/min.

3 Results and Interpretation

The axisymmetric confining pressure of the triaxial cell ($\sigma_2 = \sigma_3$) reduces the stress invariants mean effective stress p' and deviatoric stress q to

$$p' = \left(\sigma_1' + 2\sigma_3'\right) / 3 \tag{1}$$

$$q = \sigma_1' - \sigma_3' \tag{2}$$

The Bishop effective stress definition for unsaturated soils [1] describes multiphase and multi-stress geomaterials by a mechanically equivalent single phase and stress state continuum on a macroscopic scale, written in the notation of

$$\sigma' = \sigma - u_a + \chi(u_a - u_w) \quad (3)$$

where σ' is the effective stress, σ the total external stress, u_a the pore air pressure, u_w the pore water pressure, and χ is Bishop's effective stress parameter. Bishop's parameter χ is related to a volumetric function of the fluid phase to define the influence of the negative pore water pressure, and varies from 0 for the dry state to 1 for the saturated state. With the simple assumption of $\chi = S_r$, proposed by Schrefler [5], Eq. 3 can be reformulated as

$$\sigma' = \sigma - u_a + S_r(u_a - u_w) \quad (4)$$

Under saturated conditions ($S_r = 1$, $u_a = 0$) Eq. 4 reduces to Terzaghi's effective stress formulation

$$\sigma'_{sat} = \sigma - u_w \quad (5)$$

3.1 Stress–Strain and Volumetric Behaviour

In Figs. 5 and 6 the deviatoric stress q and volumetric strain ε_V are plotted over the axial strain ε_1. At unsaturated conditions a noticeable increase in peak strength is reached at lower axial compression rates throughout all tests ("suction-induced hardening"). The elastic moduli E listed in Table 1 are inferred from the secants intersecting the stress–strain curve at 0.5 of the peak deviatoric stresses q_p. Resulting E-moduli lie in the range of 10–18 MPa for the medium dense, and 30–70 MPa for the very dense soils. For the medium dense soil the matric suction of $s = 10$ kPa results in an increase of E by about 60%. For the very dense soil E remains in the range of the saturated tests.

Tests on medium dense specimens show an asymptotic convergence towards maximum deviatoric stresses under both saturated and unsaturated conditions (Fig. 5). The peak deviatoric stress of the unsaturated soil increased by about 2–5%. The unsaturated specimen consolidated to $\sigma_3 = 50$ kPa records an increase in peak strength of 19%. The volume of saturated medium dense specimens decreases with increasing axial strain (Fig. 5a). The compression rate increases with increasing consolidation pressure. The unsaturated medium dense soil specimens experience reduced contraction upon loading. The medium dense soil consolidated at $\sigma_3 = 50$ kPa transforms from contractive behaviour in the saturated state to dilative behaviour in the unsaturated state.

Stress–strain curves for the very dense specimen in Fig. 6 reach peak stresses at axial compression of $3.4\% < \varepsilon_1 < 4.8\%$, followed by strain-softening in the plastic regime under both saturated and unsaturated conditions. The peak strength of the unsaturated very dense specimen is reached at shorter axial compression in the range of $2.7\% < \varepsilon_1 < 4.2\%$. Similar to the medium dense compaction state, the peak strengths increase by about 2–5%. Experiments on saturated very dense specimens reveal strong dilative response after short initial compression phases of $\varepsilon_1 < 1\%$

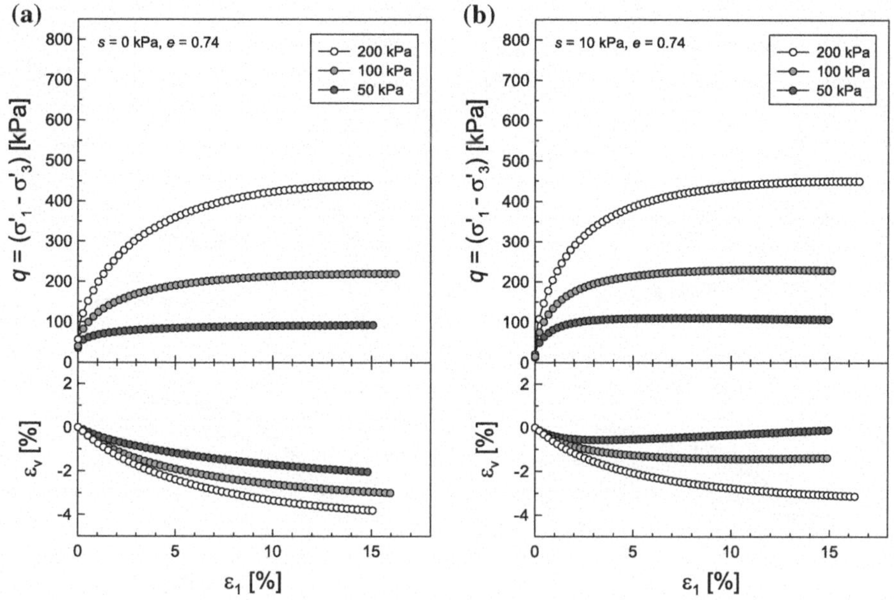

Fig. 5 Deviatoric stress and volumetric strain plotted over axial strain for the medium dense soil, under **a** saturated and **b** unsaturated conditions

(Fig. 6a). The volumetric response changes from contractive to dilative behaviour just before reaching peak stresses around axial compressions of $1\% < \varepsilon_1 < 3\%$. Smaller confining pressures result in larger dilatancy. The unsaturated soil state enhances the dilative response. The dilative rates $\dot{\varepsilon}_V$ decrease when reaching axial compressions between $5\% < \varepsilon_1 < 10\%$. The inflection points correspond with the softening regime, higher consolidation pressures result in inflection points at higher axial compressions ε_V. The shapes of the sheared soil specimen can be described by barrelling and shear failure along one distinct shear band.

Figure 7 presents the residual states of all conducted triaxial tests, resulting in a good fit for a unique critical state line for the saturated and unsaturated tests. The mean effective stress concept eliminates the influence of matric suction on the CSL [4]. The resulting critical state line has a grade coefficient $M = 1.385$ and coefficient of determination of $R^2 = 0.996$. The graphs of Fig. 6 show small decrease at residual conditions, thus corresponding points in Fig. 7 will converge towards the established critical state line.

3.2 *Shear Strength*

Figure 8 illustrates the stress paths and Mohr circles of the peak stress conditions in the σ'_{m}–τ plane, with

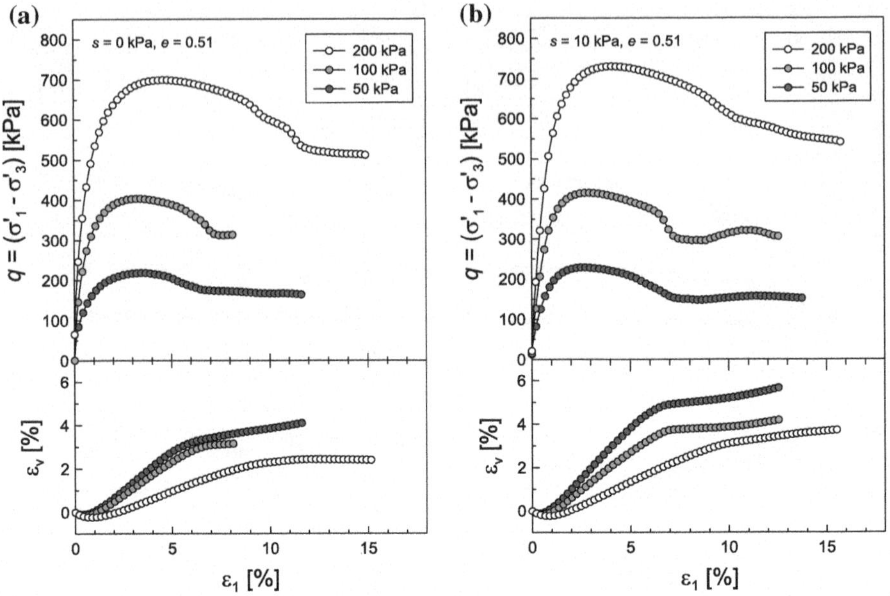

Fig. 6 Deviatoric stress and volumetric strain plotted over axial strain for the very dense soil, under **a** saturated and **b** unsaturated conditions

Fig. 7 Estimated critical state line (CSL) based on Bishop's effective stress definition

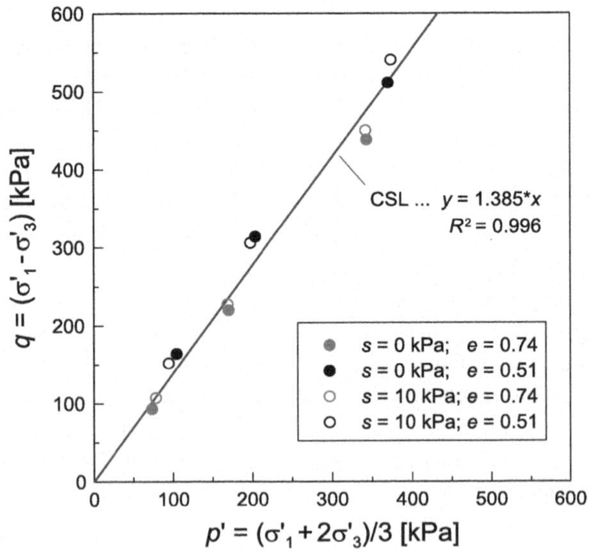

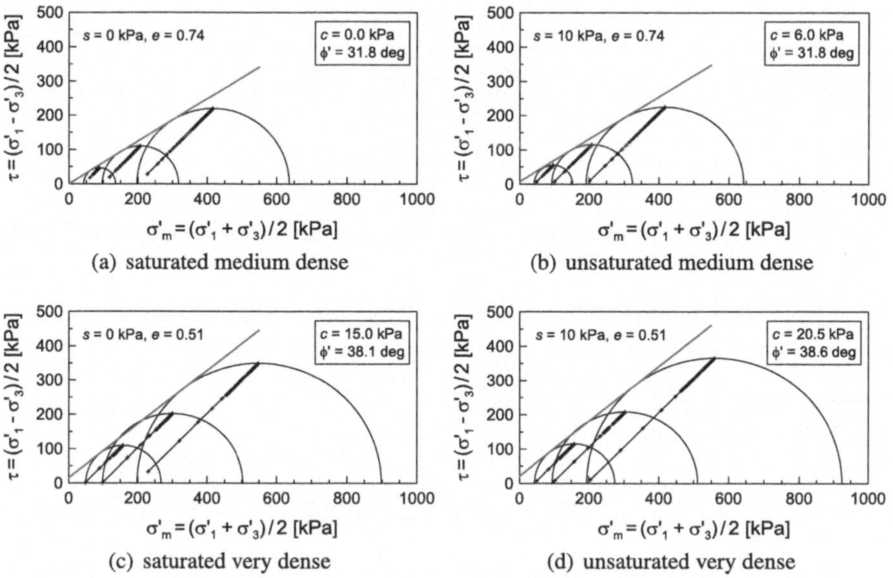

Fig. 8 Failure surface, Mohr circles for peak stresses, and stress paths of drained triaxial shear tests, with (**a**) and (**c**) saturated tests, (**b**) and (**d**) unsaturated tests with $s = 10\,\text{kPa}$

Table 2 Shear parameters obtained from drained triaxial shear tests under saturated and unsaturated conditions with $s = 10\,\text{kPa}$ for the peak and residual strength

	$e = 0.74$		$e = 0.51$			
	ϕ' (deg)	c' (kPa)	ϕ' (deg)	c' (kPa)	ϕ'_r (deg)	c'_r (kPa)
s = 0 kPa	31.8	0.0	38.1	15.0	33.0	0.0
s = 10 kPa	31.8	6.0	38.6	20.5	35.0	0.0

$$\sigma'_m = \left(\sigma'_1 + \sigma'_3\right)/2 \tag{6}$$

$$\tau = \left(\sigma'_1 - \sigma'_3\right)/2 \tag{7}$$

with σ'_m as the mean effective stress in the 2-dimensional stress space, and τ as the shear stress. The stress paths in Fig. 8 are represented by 1:1 slopes, starting at the confining pressures of 50, 100, and 200 kPa. The Mohr circles for the peak stresses indicate the stress state of failure condition. Mohr-Coulomb failure envelopes are drawn as tangential lines to the three Mohr circles of different confining pressure. The linear failure criteria show good fits for all performed test sets.

The Mohr-Coulomb shear parameters cohesion c' and friction angle ϕ' are summarized in Table 2. The examined saturated granular soil can be addressed as non-cohesive material in the medium dense compaction state, but exhibits an apparent cohesion of 15 kPa in the very dense state. The friction angle ϕ' of 31.8° for the

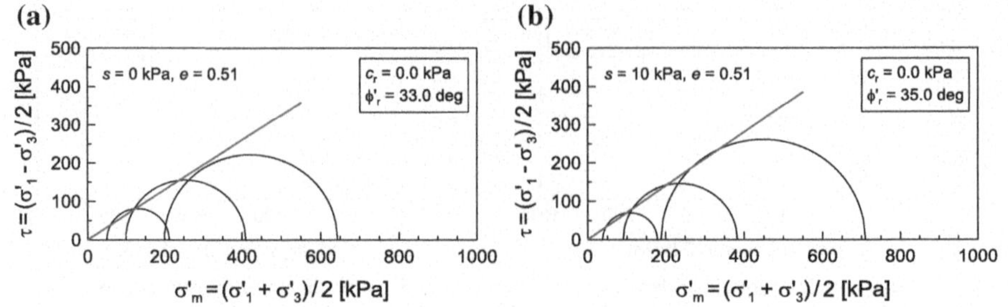

Fig. 9 Failure surface and Mohr circles for residual stresses of drained triaxial shear tests, with **a** saturated, and **b** unsaturated tests

Fig. 10 Shear angle Φ as function of confining pressure σ_3

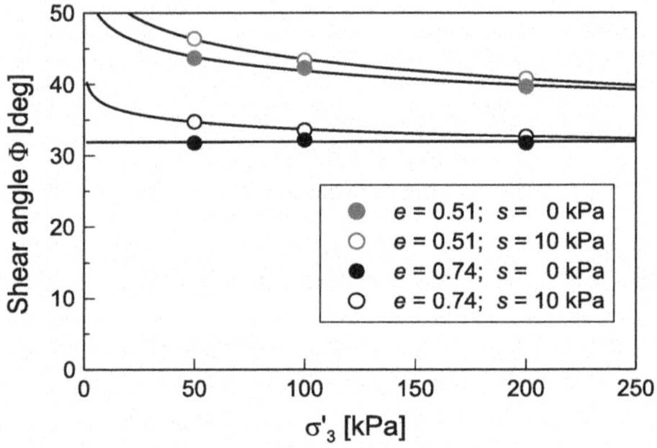

medium dense and 38.1° for the very dense compaction state is not effected by the applied matric suction of $s = 10\,\text{kPa}$, whereas the cohesion term increases from $c' = 0.0$ to $6.0\,\text{kPa}$ for the medium dense and from $c' = 15.0$ to $20.5\,\text{kPa}$ for the very dense compaction state. The additional cohesion term c'' is in the order of 6 kPa for both the medium dense and very dense configuration. The soil parameter ϕ^b [2] is close to or significantly smaller than ϕ' for the medium dense and very dense compaction state, respectively. It is noted that ϕ^b is not constant and typically decreases with higher matric suction.

The residual strength for the very dense soil is illustrated in Fig. 9. With the cohesion set equal to zero the friction angle ϕ'_r reduces to 33° and 35° for the saturated and unsaturated condition, respectively.

Figure 10 plots the shear angle Φ as function of the confining pressure σ_3. A decrease of Φ with higher σ_3 is documented for the very dense soil and the unsaturated medium dense soil, whereas Φ of the saturated medium dense soil is unaffected by the confining pressure. The unsaturated soil state results in higher shear angles. The highest increase of shear angles is observed for the lowest confining pressure of 50 kPa.

4 Conclusions

In this paper, the significant contribution of capillarity to the mechanical behaviour and shear strength of the unsaturated granular soil under shear loading has been shown in the experimental investigation. The unsaturated soils with matric suction of $s = 10\,\text{kPa}$ show increase in peak stress at lower axial compression for both medium dense and very dense compaction states, a behaviour known as "suction-induced hardening". The applied matric suction reduces the contractive response upon loading, and can even result in dilative response of the granular soil. The increase in shear strength of the granular soil is expressed by an additional term of apparent cohesion. The applied matric suction increases the cohesion term by about 6 kPa, whereas the friction angle is not significantly affected by the unsaturated state under the examined condition. The additional strength from capillarity is most pronounced for the lowest investigated confining pressure.

Acknowledgements The support of the Otto Pregl Stiftung is greatly appreciated.

References

1. Bishop, A.W.: The principle of effective stress. Tecnisk Ukeblad **39**, 859–863 (1959)
2. Fredlund, D.G., Morgenstern, N.R., Widger, R.A.: The shear strength of unsaturated soils. Can. Geotech. J. **15**, 313–321 (1978)
3. Idinger, G.: Experimental study of failure initiation in partially saturated slopes. Ph.D. thesis, Universität für Bodenkultur, Wien (2016)
4. Nuth, M., Laloui, L.: Effective stress concept in unsaturated soils: clarification and validation of a unified framework. Int. J. Numer. Anal. Meth. Geomech. **32**(7), 771–801 (2008)
5. Schrefler, B.A.: The finite element method in soil consolidation (with applications to surface subsidence). Ph.D. thesis, University College of Swansea (1984)
6. Wheeler, S.J.: The stress-strain behavior of soils containing gas bubbles. Ph.D. thesis, The University of Oxford (1986)

Failure of Unsaturated Soil Slopes Initiated by Self-weight Loading

Gregor Idinger and Wei Wu

Abstract A set of centrifuge model tests is performed to investigate the stability and failure initiation of unsaturated soil slopes. Homogeneous model slopes of reduced size are tested in the enhanced acceleration field of the geotechnical centrifuge. The body stress increases due to increasing self-weight until failure conditions are reached. The pre-failure soil displacement and the propagation of the shear zone is visualized by the digital image correlation technique. The experimental parametric study with granular soil investigates the role of moisture content, soil density, and slope inclination. Observed failure modes can be classified as shallow translational slides of coherent soil blocks of low mobility. The shear zone emerges at regions of local weakness, and is observed to initiate at the toe of the slope or around mid-height of the slope.

1 Introduction

The small number of field observations on the triggering of slope failure is evidence of the difficulties associated with the monitoring of the pre-failure condition. In addition to financial and manpower expenses there remains the challenge of ensuring reliable long-term in-situ measurements. An exhaustive study requires accurate measurements of initial soil deformation and the moisture content of the soil. Yet, in-situ monitoring networks have declined since the 1980s despite the growing interest in slope failure mechanisms and landslides due to the rise of complementary

G. Idinger (✉) · W. Wu
Institute of Geotechnical Engineering, University of Natural Resources and Life Sciences, Feistmantelstrasse 4, 1180 Vienna, Austria
e-mail: gregor.idinger@boku.ac.at

W. Wu
e-mail: wei.wu@boku.ac.at

W. Wu (ed.), *Recent Advances in Geotechnical Research*,
Springer Series in Geomechanics and Geoengineering,
https://doi.org/10.1007/978-3-319-89671-7_8

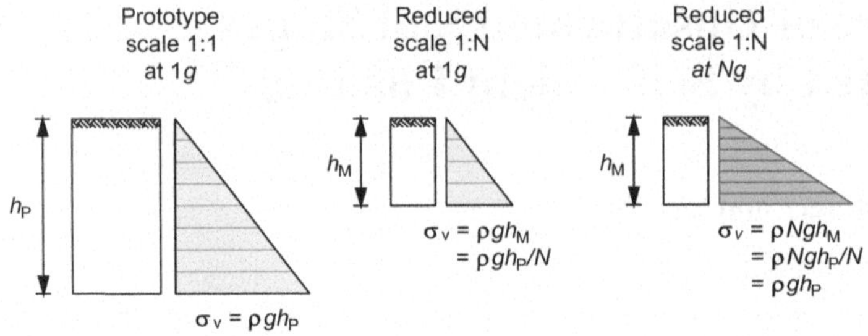

Fig. 1 Schematic presentation of the principle of centrifuge modelling, with the profiles of vertical stress for the prototype, small-scale model, and the small scale model exposed to centrifuge acceleration

satellite remote sensing [1]. In lieu of in-situ measurements the back analysis with limit equilibrium models is often the only possibility to evaluate the failure condition and failure mechanism of documented slope failures, e.g., by Iverson et al. [4]. The complexity of the spatially and temporally varying underground condition complicates any potential analysis of observed behaviour of field slopes. The gained data and conclusions of failure mechanism are often limited to the investigated case study. Hence, the direct calibration of numerical models on field observations can be dangerously misleading as field data are generally subject to several uncertainties on actual ground conditions.

Physical model tests on slopes of simplified geometry and strata allow to study the failure mechanism under controllable boundary conditions and known material properties, thus can provide valuable data for the calibration and validation of numerical models. For physical slope models of reduced-size it is crucial to replicate the correct mechanical behaviour of the soil. As the strength of soils varies dramatically with the stress level it is crucial to replicate the stress state of the full-scale field situation (prototype) properly. This can be achieved with the centrifuge model technique, as illustrated in Fig. 1.

The introduced set of physical model tests in the geotechnical centrifuge aims (i) to describe pre-failure deformations and to localize the initiation of shear zones, (ii) to describe the failure mode, and (iii) to find the maximum slope height for the prototype scale that withstands failure (critical slope height) at various soil conditions.

2 Material and methods

All centrifuge tests are performed at the centrifuge testing facility of the Institute of Geotechnical Engineering at the University of Natural Resources and Life Sciences in Vienna. Figure 2 presents a schematic illustration of the beam-type BOKU-IGT centrifuge. The radius from the axis of rotation to the swinging platform is 1.30 m.

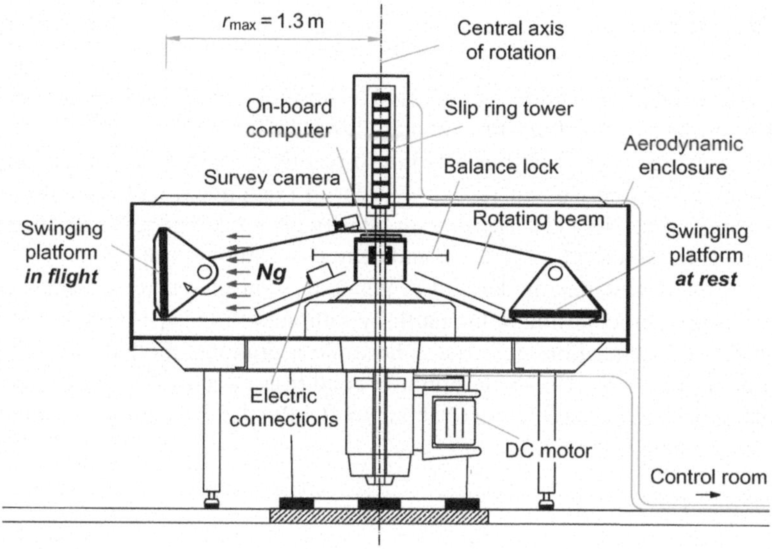

Fig. 2 Schematic presentation of the BOKU-IGT centrifuge testing facility, model '1231 Standard Heavy Duty' manufactured by *Trio-Tech*

Table 1 Properties of silty medium-fine sand

Property	Value
e_{max}	1.09
e_{min}	0.45
d_{50} (mm)	0.23
d_{10} (mm)	0.022
C_c ($d_{30}^2/d_{10}d_{60}$)	3.6
C_u (d_{60}/d_{10})	13.6
ρ_s (g/cm^3)	2.67
w_{opt}	15.1%
Friction angle* ϕ'	31.8°
Cohesion* c' [kPa]	6.0

*Determined for $\rho_d = 1.53$ g/cm^3 and $w = 11.5\%$

The maximum load capacity is 10 g-t, e.g. 100 kg at 100 g. The size of models is limited to 540 × 560 × 560 mm ($w \times d \times h$).

The model slope is built and tested in a plane-strain container of aluminium rear and side walls of 15 mm thickness and transparent acrylic front wall of 30 mm thickness. The transparent front wall is used as an observation window, revealing the side profile of the plane-strain model slope. A thin film of clear colourless silicon oil establishes a low-frictional boundary on the inside walls, but does not effect the DIC analysis on soil deformation.

2.1 Model Soil

The experimental investigation is conducted on unsaturated granular model soil. Key properties of the silty medium-fine sand are listed in Table 1. The narrow grain size distribution curve (~80% within 0.1–1 mm) results in narrow transition zones of the soil water characteristic curve of one order of magnitude, with air-entry values ranging from 3.5–6.0 kPa [2]. The Proctor density $\rho_{Pr} = 1.805$ g/cm^3 is reached at the optimum water content of $w_{opt} = 15$ % and degree of saturation of 0.84.

The saturated granular model soil is generally considered to be cohesionless, yet reveals apparent cohesion in the partially saturated state [3]. At the investigated medium dense compaction state the cohesionless granular model soil features a friction angle of 31.8°. Under unsaturated condition (water content of $w = 11.5\%$ with corresponding matric suction of about $s = 10$ kPa) the soil exhibits an additional apparent cohesion of $c' = 6$ kPa.

2.2 Instrumentation

A digital still camera records the pre-failure deformation of the visible side profile of the model slope through the plane-strain container's transparent front wall. The domain of interest of the deformation analysis is highlighted in Fig. 3. During the post-failure stage a high-speed video camera allows to track the mobilized soil mass. The analysis of pre-failure slope deformation is performed with the programme *GeoPIV8*, developed by Take and White [8]. *GeoPIV8* allows micro-mechanical observations prior failure at any point of the visible soil model, including soil displacement and velocity, volumetric strain, and shear strain at sub-pixel precision. The digital image correlation (DIC) technique enables the contact-less and non-disturbing tracking of the soil surface using the inherent random texture of the medium-fine sand.

Penetrometer tests with a miniaturized probe give qualitative information on the compaction effort and homogeneity of the layer-wise compaction over the model height, especially within layers (intra-layer homogeneity). Penetrometer tests are performed on the initial model soil and after self-weight compaction in the centrifuge. The water content is determined according to the oven-drying procedure on small soil samples taken with a small thin-walled cylindrical probe with diameter of $d = 20$ mm and height of $h = 20$ mm on the initial state and post-test on the failure state.

2.3 Experimental Programme

Figure 3 illustrates the geometry of the investigated homogeneous soil slopes. Model slopes have a height of 200 mm and are located on a base layer of 50 mm height. The plane-strain width of the model is 155 mm. The slope crest has a length of 180 mm.

Model slopes are inclined at 45° and 55°, with the angle of slope inclination β defined above the horizontal. The model is compacted by moist-tamping in 5 layers of 50 mm, then cut to the desired slope geometry using lateral templates. The boundaries along the sides and bottom of the slope provide no-flow conditions.

The failure of model slopes is evoked by the slow continuous increase of the gravity-induced stress field (Fig. 4). The centrifugal acceleration is increased at the

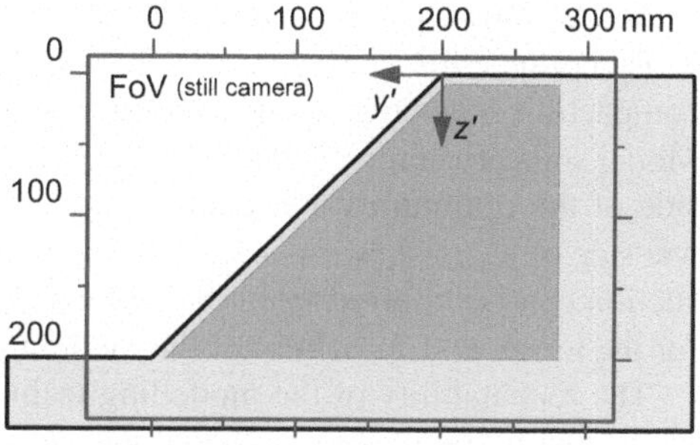

Fig. 3 Schematic drawing of the homogeneous slope with the camera's field of view and the domain of deformation analysis highlighted

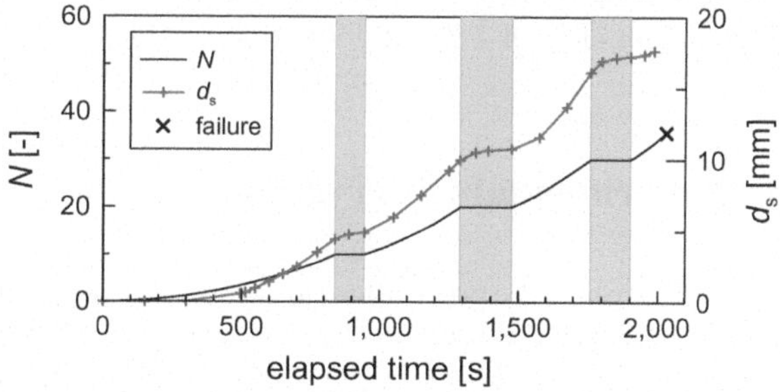

Fig. 4 Observed model response during gravity turn-on until slope failure by the recorded surface settlement d_s of the crest of the slope due to self-weight compaction (GUP-09)

Table 2 Experimental programme of the centrifuge test series on self-weight loading.

Test ID	Initial conditions			Failure conditions					
	β	$\rho_{d,0}$	w_0	$\rho_{d,f}$	w_f	max g-level		$h_{f,P}$	Failure
	(deg)	(g/cm^3)	(g/cm^3)	(%)	(%)	(rpm)	(g)	(m)	(–)
GUP-01	55	1.40	11.5	1.41	11.4	126.0	22.0	4.4	Y
GUP-02	55	1.45	11.5	1.46	11.4	167.0	38.6	7.7	Y
GUP-03	55	1.45	11.5	1.46	11.3	168.5	39.3	7.9	Y
GUP-04	55	1.50	11.3	1.51	11.2	204.1	57.7	11.5	Y
GUP-05	55	1.53	11.3	1.54	11.2	209.6	60.8	12.2	Y
GUP-06	45	1.45	11.1	1.51	11.0	270.0	100.9	–	N
GUP-07	55	1.45	17.1	1.53	14.7	188.6	49.2	9.8	Y
GUP-08	55	1.53	17.3	1.59	13.3	268.0	99.4	–	N
GUP-09	55	1.45	17.4	1.56	16.5	160.7	35.7	7.1	Y
GUP-10	55	1.53	18.0	1.60	16.4	200.2	55.5	11.1	Y

rate of 1 rpm in 10 s. The stress increase can be interpreted as an increase of the prototype's slope height while maintaining the slope geometry [7]. The maximum slope height of the prototype which provides stable conditions can be inferred from the centrifugal acceleration prior failure initiation, referred to as the critical slope height. Tests are terminated after slope failure or when reaching an acceleration of 100 g, which corresponds to the prototype slope height of 20 metres. In the event of slope failure Table 2, which summarizes the experimental programme, lists the critical slope height for the prototype scale $h_{f,P}$.

Model soils are compacted to loose, i.e., $D_r = 0.29$ ($\rho_d = 1.40\,g/cm^3$), or medium dense compaction states, i.e., $D_r = 0.40$, 0.49, 0.55 ($\rho_d = 1.45$, 1.50, 1.53). These compaction states represent collapsible and naturally weathered soils, respectively. Model soils are compacted at two different water contents, corresponding to the dry side of the optimum water content w_{opt}, i.e., $w_0 < w_{opt}$ with $w_0 = 11.5\%$, or to the wet side of w_{opt}, i.e., $w_0 > w_{opt}$ with $w_0 = 18.0\%$, respectively. The associated matric suction can be inferred from the soil water characteristic curve, and is about 6 kPa for the loose, and 7–10 kPa for the medium dense soil [2].

The repeatability of the modelling technique for the investigated slope model is confirmed by two tests conducted with identical initial conditions (GUP-02, GUP-03). Both slopes failed at almost identical g-levels ($N = 38.6$, 39.3) related to critical slope heights $h_{f,P}$ of 7.7 and 7.9 m.

3 Experimental Observations

Figure 5 displays the post-failure appearance of slopes inclined at 55° and compacted at $w < w_{opt}$ (GUP-04, GUP-05) or $w > w_{opt}$ (GUP-09). The mobilized soil masses of similar volume deposit after short sliding paths under an angle of about 45°. The debris mass from slopes compacted at $w > w_{opt}$ is deposited at smaller extends and therefore at slightly higher deposition angles. As illustrated in Fig. 5, tension cracks may appear at the top of the slope prior the sliding event, especially for slopes compacted at $w < w_{opt}$. In case of several tension cracks, the first arises at the furthermost distance to the slope crest, with subsequent tension cracks appearing successively closer to the crest of the slope. Finally, the slip surface exits through the tension crack wich is at the closest to the slope crest. Tension cracks in vicinity to the tearing edge indicate the unstable nature of the remaining steep slope section, as displayed in Fig. 5. Under a further increase in g-level those unstable soil blocks tend to fail.

Role of slope inclination

While homogeneous soil slopes inclined at 55° fail under the self-weight loading condition at g-levels ranging from 22 to 61 g, the slope inclined at 45° (GUP-06) remains stable up to 100 g and does not show any significant shear deformation. Thus, the steeper slope angle decreases the stability of the slope and results in smaller critical slope heights.

Role of void ratio

In the centrifugal acceleration field the model soil experiences compaction due to its enhanced self-weight. This self-weight compaction is more pronounced in soils compacted at the high water content of $w > w_{opt}$. The dry densities of the failure state $\rho_{d,f}$ are back-calculated from the initial dry density and the recorded settlement. The image correlation analysis reveals homogeneous compaction throughout the height of the slope, which ensures homogeneous soil density during the failure stage, as demonstrated in [2]. The dry density increases by 0.8–1.1% for soils compacted at $w < w_{opt}$ and 4.8–7.3% for soils compacted at $w > w_{opt}$, with higher figures related to lower compaction ratios.

Slopes compacted to higher dry density withstand higher g-levels, as presented in Fig. 6. Figure 6 includes the initial dry density and the increased dry density representing the condition just before slope failure. The fitting curves connecting the points representing slopes compacted at the same water content show linear corre-

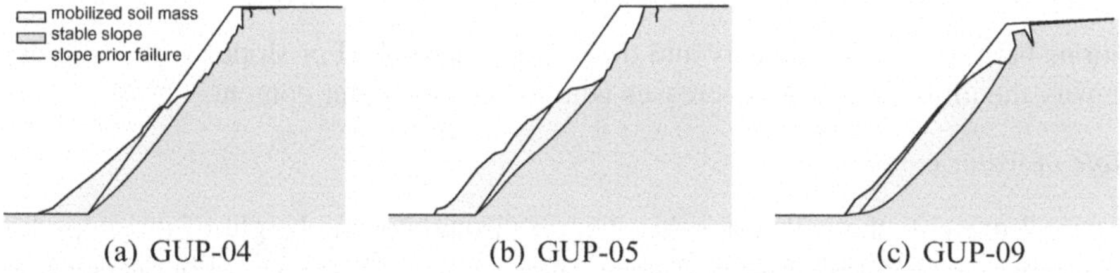

Fig. 5 Post-failure appearance with deposition and volume of the mobilized soil mass, for slopes compacted at $w < w_{opt}$ (GUP-04, GUP-05) and $w > w_{opt}$ (GUP-09). The shape of the slope prior failure is indicated by the thin black line

Fig. 6 The g-level at slope failure, as function of dry density and water content. The empty points represent the initial dry density, the filled points represent the dry density at failure condition after self-weight compaction of the same test

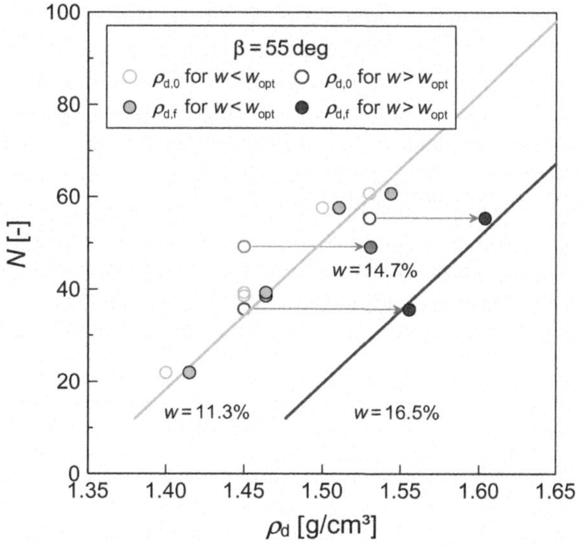

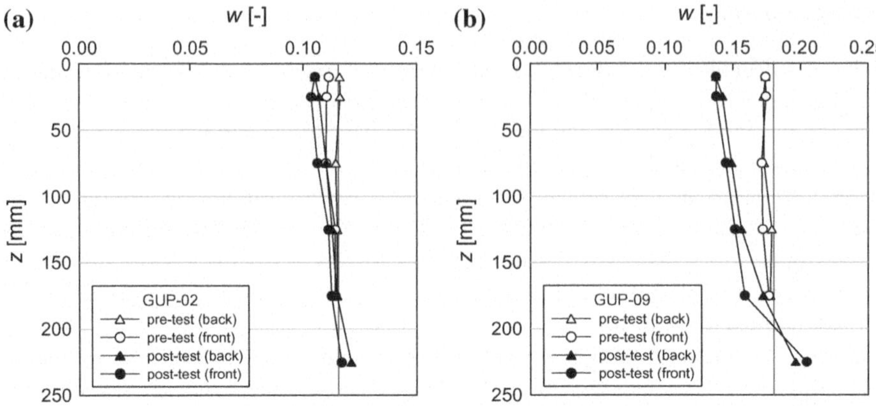

Fig. 7 Profiles of soil water over the model height measured after soil compaction and directly after the centrifuge test, for a slope compactes at **a** $w < w_{opt}$ (GUP-02) and **b** $w > w_{opt}$ GUP-09

lations between the dry density and the scaling factor N. For slopes with same dry density the ultimate g-level decreases with increasing water content.

Role of water content

Figure 7 presents the soil-water distribution throughout the height of the slope for representative slopes with water contents of $w < w_{opt}$ and $w > w_{opt}$. For the determination of the water content small soil samples are taken along the transparent front wall and the back wall of the plane-strain model slope over the height of the slope. White symbols represent the initial state after model preparation w_0, black symbols represent the water content related to failure condition w_f. The targeted water content is indicated by the vertical line. The compaction layers are indicated by horizontal lines. Upon spin-up the migration of soil water is obvious from the presented profiles of water content. The gravitational flow towards the bottom layer is more pronounced for slopes with the high initial water content $w > w_{opt}$ (Fig. 7b). This can be explained by the different micro-structure with large inter-aggregate pores observed for soils compacted at $w > w_{opt}$, which corresponds to degrees of saturation of 0.6–0.7, resulting in a higher hydraulic conductivity. The average difference in w_f from soil samples taken from the front and back side for soils compacted at $w < w_{opt}$ shows a difference of 0.2%, for soils compacted at $w > w_{opt}$ a difference of 0.7%. Thus, the radial centrifugal component has minor influence on the migration of soil water in the eccentrically positioned model slope.

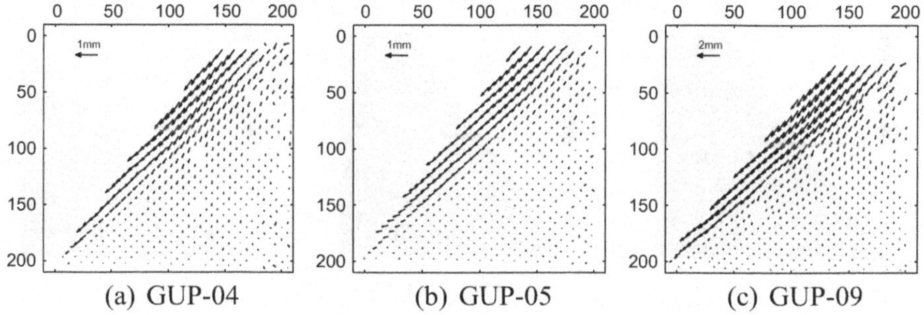

Fig. 8 DIC result on the soil deformation within the last 50 s before failure, for slopes compacted at $w < w_{opt}$ (GUP-04, GUP-05) and $w > w_{opt}$ (GUP-09, GUP-10)

3.1 Pre-failure Deformation

The pre-failure displacement of mobilised sliding masses is shown in the vector graphs of Fig. 8, presenting the total soil deformation within the last 50 s prior failure. The presented slopes are inclined at 55° and compacted to medium dense compaction states at water contents of $w < w_{opt}$ (GUP-04, GUP-05) or $w > w_{opt}$ (GUP-09). All slopes show similar patterns of pre-failure deformation. The displacement vectors of the destabilized soil mass are oriented parallel down-slope at similar magnitudes inclined at principal direction of 45°, almost parallel to the sliding plane. This observation indicates the shallow translational failure of coherent soil blocks.

In order to isolate the pre-failure sliding deformation from the the total soil deformation Figs. 9 and 10 present the horizontal component of the pre-failure deformation y' over increasing g-level, for representative slopes compacted at $w < w_{opt}$ (GUP-04) and at $w > w_{opt}$ (GUP-09). The mobilized soil mass of slopes compacted at $w < w_{opt}$ shows smaller extent in depth than slopes compacted at the high water content. The average height of the mobilized soil mass is around 30 mm. While slopes compacted at $w < w_{opt}$ show peak horizontal displacements at shallow depths throughout the full slope height (Fig. 9), slopes compacted at $w > w_{opt}$ show peak horizontal deformations around the toe of the slope (Fig. 10). In combination with higher vertical deformations at the crest region the failure mechanism of slopes compacted at $w > w_{opt}$ exhibit rotational tendency.

During the failure stage the mobilized soil mass exhibits maximum velocities of about 0.6 m/s. It is noted that velocity for dynamic and inertial events are the same in the model and the respective prototype [7].

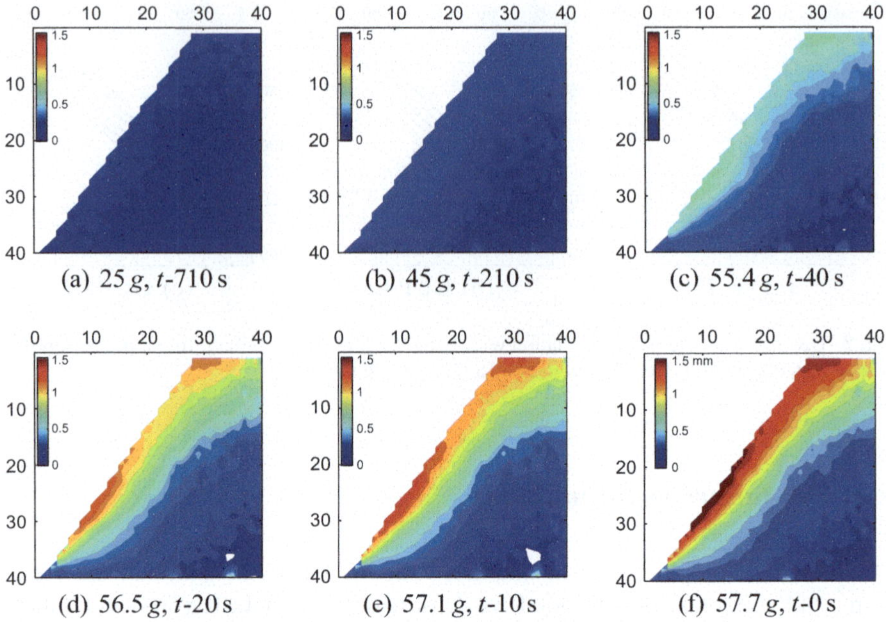

Fig. 9 Horizontal y' component of pre-failure soil displacement in test GUP-04

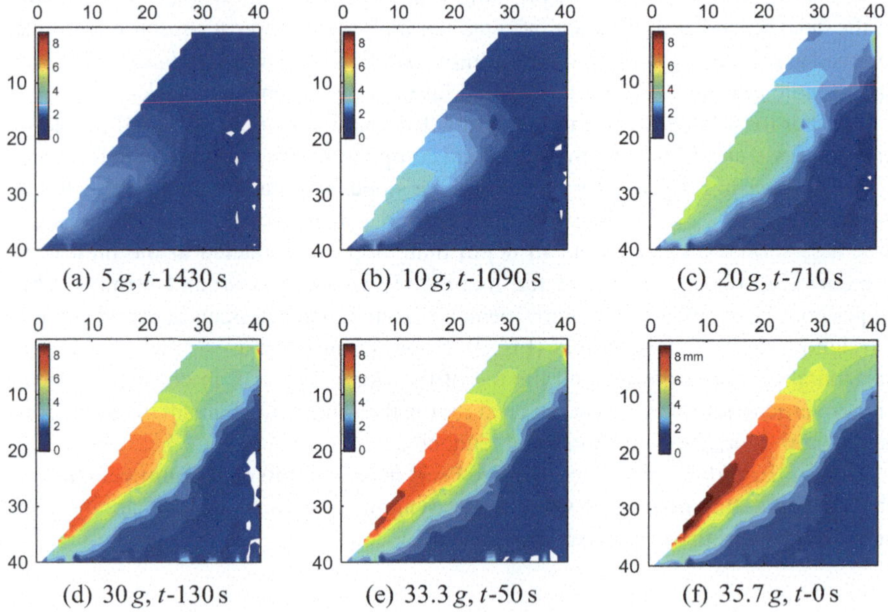

Fig. 10 Horizontal y' component of pre-failure soil displacement in test GUP-09

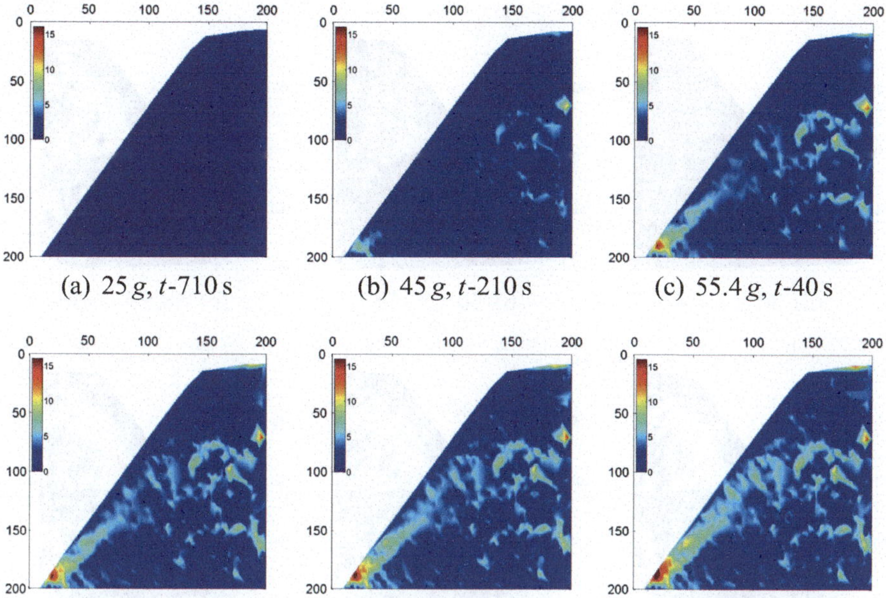

(a) 25 g, t-710 s (b) 45 g, t-210 s (c) 55.4 g, t-40 s

(d) 56.5 g, t-20 s (e) 57.1 g, t-10 s (f) 57.7 g, t-0 s

Fig. 11 Development of the shear zone initiating from the toe of the slope, as in test GUP-04

3.2 Shear Deformation

Figures 11 and 12 illustrate the cumulative propagation of the shear strain for two slopes of similar initial condition. The progressive failure is presented in terms of g-level and time t counting towards failure. As in the majority of tests a band of high shear strains initiates at the toe of the slope and propagates almost linear under an angle of $45 \pm 2°$, as indicated in Fig. 11. In slopes compacted at $w > w_{\mathrm{opt}}$ the shear band is inclined at the values from the lower range. Slopes compacted at $w > w_{\mathrm{opt}}$ experience noticeable compaction along the shear zone. Alternatively, the shear zone is observed to emerge at the mid-height of the slope, with shear strains increasing rather simultaneously within the shear band, as displayed in Fig. 12. This second failure type is comparable to findings published by Lade [5], who concluded that the region of instability emerges at the mid-height of the slope where the shear stress is at maximum. The emerging of the first shear strains can be linked to local weak zones, which are apparent in the less compacted lowermost region of compaction layers. As a consequence, the failure process is at first local with the formation of localized plastic zones, then propagates to global failure. It is noted that the initiation of shear deformation is detected at g-levels slightly lower than the g-level at which the slope fails. For instance, in test GUP-04 the scaling factor N increased by 2.3 g within the last 40 s prior failure, the period represented by Fig. 11c–f.

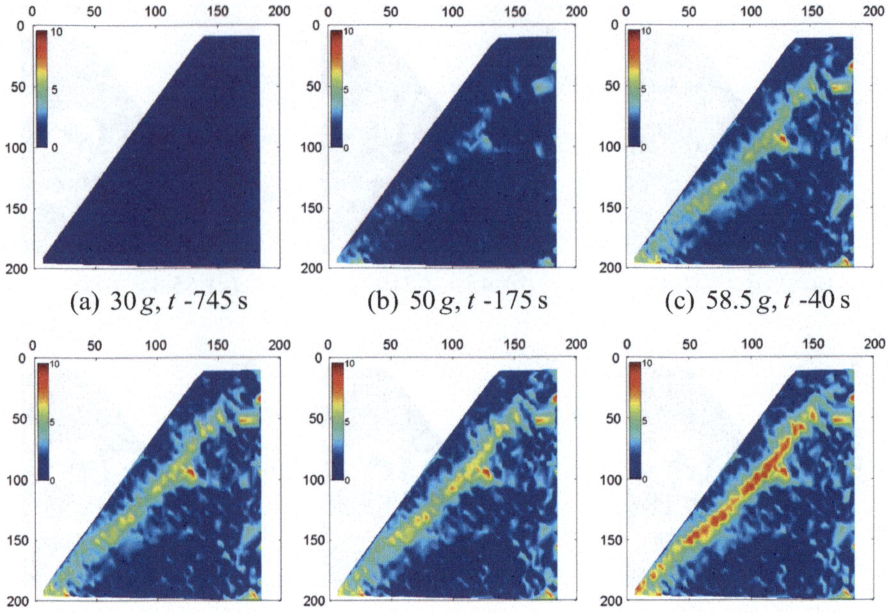

(a) 30 g, t -745 s (b) 50 g, t -175 s (c) 58.5 g, t -40 s

(d) 59.7 g, t -20 s (e) 60.2 g, t -10 s (f) 60.8 g, t 0 s

Fig. 12 Development of the shear zone initiating at mid-height of the slope, as in test GUP-05

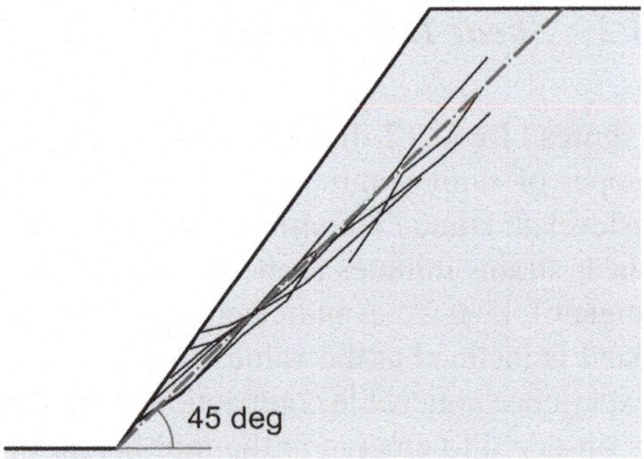

Fig. 13 Superposition of observed failure planes for slopes inclined at 55°, in relation to a plane inclined at 45°

All recorded shear zones from slopes inclined at 55° are superposed in Fig. 13. It is shown that the sliding planes are inclined under an angle of about 45°. This could explain why the slope inclined at 45° has not failed.

The width of the shear band in granular materials is related to the mean grain diameter of the soil [6]. In centrifuge models the width of shear bands appear at the prototype scale, as the size of soil particles remain unscaled. Recorded shear zones have widths ranging from 9 to 12 mm, which corresponds to 40–50 times the mean grain diameter d_{50}. This is close to the lower bound of the bifurcation strain for fine sand as defined in [6].

4 Conclusions

The presented series of centrifuge model tests investigates the failure initiation of unsaturated granular soil slopes. The model slopes are exposed to increasing body stresses to trigger slope failure. The stability and failure mode of unsaturated soil slopes is influenced by the slope inclination, degree of saturation, and soil density of the granular model soil. The maximum slope height of the prototype slopes is found to follow a linear trend with increasing dry density. The ultimate slope height at the prototype scale of the investigated model slopes ranges from 4.4 to 12.2 m. The failure mode can be classified as non-liquefied, shallow translational slides of low mobility. The failed soil mass rests on the base of the slope after short travel distances. In slopes compacted at $w > w_{\mathrm{opt}}$ the deeper failure plane mobilises larger soil volumes. In soils with the high water content $w > w_{\mathrm{opt}}$ a gravitational flow of soil water is observed towards the bottom of the model during spin-up. The shear zones initiate at regions of local weakness, commonly observed to emerge at the toe of the slope, with the shear zone propagating linearly towards the top of the slope under an angle of about 45°.

Acknowledgements The support of the Otto Pregl Stiftung and the HORIZON 2020 project Geohazards Risk Assessment, Mitigation and Prevention (GEO-RAMP, No. 645665) GEO-RAMP project is greatly appreciated.

References

1. Fekete, B.B.M., Robarts, R.D., Kumagai, M., Nachtnebel, H.-P., Odada, E., Zhulidov, A.V.: Time for in situ renaissance. Science **349**(6249), 685–686 (2015)
2. Idinger, G.: Experimental study of failure initiation in partially saturated slopes. Ph.D. thesis, Universität für Bodenkultur, Wien (2016)
3. Idinger, G., Wu, W.: Shear strength of granular soil under saturated and unsaturated conditions. In: Recent Advances in Geotechnical Research, pp. 79–89. Springer, Berlin (2019)
4. Iverson, R.M., George, D.L., Allstadt, K., Reid, M.E., Collins, B.D., Vallance, J.W., Schilling, S.P., Godt, J.W., Cannon, C.M., Magirl, C.S., Baum, R.L., Coe, J.A., Schulz, W.H., Bower, J.B.: Landslide mobility and hazards: implications of the 2014 Oso disaster. Earth Planet. Sci. Lett. **412**, 197–208 (2015)
5. Lade, P.V.: Static instability and liquefaction of loose fine sandy slopes. J. Geotech. Eng. **118**(1), 51–71 (1992)
6. Mühlhaus, H.-B., Vardoulakis, I.: Thickness of shear bands in granular materials. Géotechnique **37**(3), 271–283 (1987)
7. Taylor, R.N.: Geotechnical Centrifuge Technology. Taylor & Francis, London (1995)
8. White, D.J., Take, W.A., Bolton, M.D.: Soil deformation measurement using particle image velocimetry (PIV) and photogrammetry. Géotechnique **53**(7), 619–631 (2003)

Centrifuge Study of Soil Arching in Slope Reinforced by Piles

Guoping Lei, Sara Usai and Wei Wu

Abstract The main mechanism of discretely installed stabilizing piles is soil arching, which allows the driving force exerted on the soil mass between the piles to be transferred to the piles. This paper presents the results of a series of centrifuge tests of a 90° slope reinforced by a row of piles with different pile spacing. The soil can move freely between the piles. Based on the failing of soil mass, the soil stabilized by the piles was divided into three zones, in which the most important soil area was drawn. Two failure modes of soil arching was observed in the tests. The soil arch trends to fail at the dome when the pile spacing is large, and at the pile-soil contact when pile spacing is small.

1 Introduction

Soil arching, as described by Terzaghi [17], is one of the most universal phenomena encountered in soils both in the field and in the laboratory. Through soil arching, the soil stresses from a yielding part can be transferred to adjacent stationary parts. During the past decades, it has been studied in various engineering practice: Tunnelling [10], piled embankments ([1, 6, 13, 18], etc.), retaining wall ([5, 14, 15], etc.), piled slope

G. Lei (✉) · W. Wu
Institute of Geotechnical Engineering,
University of Natural Resources and Life Sciences, Vienna,
Feistmantelstrasse 4, 1180 Wienna, Austria
e-mail: guoping.lei@students.boku.ac.at

W. Wu
e-mail: wei.wu@boku.ac.at

S. Usai
Faculty of Engineering and Architecture, University of Cagliari,
Via Marengo 2, 09100 Cagliari, Italy
e-mail: saracala@hotmail.it

W. Wu (ed.), *Recent Advances in Geotechnical Research*,
Springer Series in Geomechanics and Geoengineering,
https://doi.org/10.1007/978-3-319-89671-7_9

([3, 11, 12, 19], etc), etc. It can also be drawn from the literature that among the above-mentioned engineering topics the soil arching in piled embankments has the highest level of understanding. Various analytical methods have been developed to model the arching effect and calculate the stress acting on the structure. Good agreement has been obtained in the comparison between analytical results and observed data. The successful application of soil arching theory in this topic benefited from the simple initial and boundary conditions. Similar to the yielding trap door experiment, in piled embankments, the initial stress field is determined by static gravity and no initial shear stress exists. For the boundary conditions, the stationary support adjoining the trap door and the piles in piled embankments are reasonably simplified as rigid. Unfortunately, neither of the initial and boundary condition is valid for stabilizing piles. The driving force exerted on the piles, which originally should be transferred to the soil slice at the same location, depends on the relative pile-soil movement. The piles, which must penetrate through the sliding soil mass into a firm stable strata, usually work as flexible cantilever beams. Consequently, the driving force usually causes further non-uniform displacement, which in return affects the deformation of the failing soil mass and therefore the driving force itself. This complex coupled problem must be solved simultaneously. Wang and Yen [19] applied Terzaghi's shear plane method [17] to an infinite piled slope with the assumption of rigid-plastic soil above a firm underlying base. The theory had made great progress to estimate the arching effect in piled slope, but the limitation was also apparent due to the oversimplifying of the geometry and soil properties. Ito and Matsui [7] proposed methods to estimate the forces acting on the piles but also based on the assumption of plastic state soil. Bosscher and Gray [3] performed laboratory tests on soil arching in sandy slopes and found out that the closer the pile spacing the stronger the arching effect. The piles were simulated by a swing gate with certain width. Liang and Zeng [12] conducted two-dimensional finite element analyses of a horizontal soil-pile slice which was subjected to in-plane soil movements. Parameters referring to soil arching were examined and it was concluded that the ratio of pile spacing to pile diameter was the most important design factor impacting both the possibility and intensity of soil arching. Chen and Martin [4] linked the arching effect to pile load-displacement curves obtained from two-dimensional finite difference analyses. Various pile and soil parameters were studied, in which the consideration of drained and undrained conditions, the effect of interface roughness and soil dilatancy and the comparison between active and passive pile response are notable. Li et al. [11] also conducted numerical tests on a horizontal soil-pile slice but subjected to horizontal thrust at the boundary of the numerical model. The thrust was found to be shared by three parts: soil arching behind the piles, soil arching between the piles and sliding mass in front of the piles. The three-dimensional arching effect for piles under lateral soil movement is far more complicated and is usually indirectly investigated by comparing the pile response in different pile spacing occasions undergoing the same soil movement ([8, 9, 16], etc.). The objective of this study is to model the soil arching in a piled slope in a centrifuge and analyse its role in slope stability.

2 Centrifuge Modelling

As illustrated in Fig. 1, a 90° slope reinforced by a row of discrete piles is modelled here. The number of the piles varies in different tests. The model should represent an infinite slope along the pile row, therefore the boundaries need to be the symmetry plane of the infinite pile row, which is either at the middle between two adjacent piles or at the axis of a pile. The former is adopted in this study in order to capture the arching effect from the exposed side boundary. No restriction is applied between the piles. The model soil was medium sand with a density of 1.53g/cm^3 and a water content of 11.5%. The model piles were made of aluminium tubes with an outer diameter of 10 mm and a wall thickness of 1 mm. The length of the piles is 240 mm and they were embedded into the firm ground for 110 mm. Generally, the model was constructed in three steps: (1) The compaction of the mixed soil via moist tamping; (2) The installation of piles; (3) The cutting of the slope to prescribed geometry. In principle there should be no shearing stress at the side boundaries since they are the symmetry cross sections. This requires the friction provided by the side walls to be minimised as much as possible. Here silicon oil was applied on the side walls during the construction.

Since the only reason that triggers the failure of the piled slope is gravity, G-up test procedure, which means continuous increase of the gravity-induced stress field, is employed. The stress increase in the model slope corresponds to the increase of prototype slope height. The g-up tests are designed to end when the slope-pile system fails or 50 g is reached. The increase of centrifugal acceleration is conducted at a low rate of 0.1 rpm per second by a computer program. In-flight photography both from the side and the top was used to monitor the movements of the slope. PIV analysis [20] was employed to quantitatively study the soil deformation.

3 Results

Results of 5 centrifuge tests with different pile spacing (namely s/d = 3.88, 5.17, 7.75, 15.5 and one test without piles) are presented as follows.

3.1 Unreinforced Slope

Figure 2 shows the side view before and after the failure of the slope without piles and Fig. 3 shows the corresponding top view. A tension crack at the crest was observed before the failure. Along the direction perpendicular to the cross sectional plane, the tension crack has a slight arching shape because of the friction of the back wall and the front acrylic plate. The friction effect, which is unavoidable, was reduced by

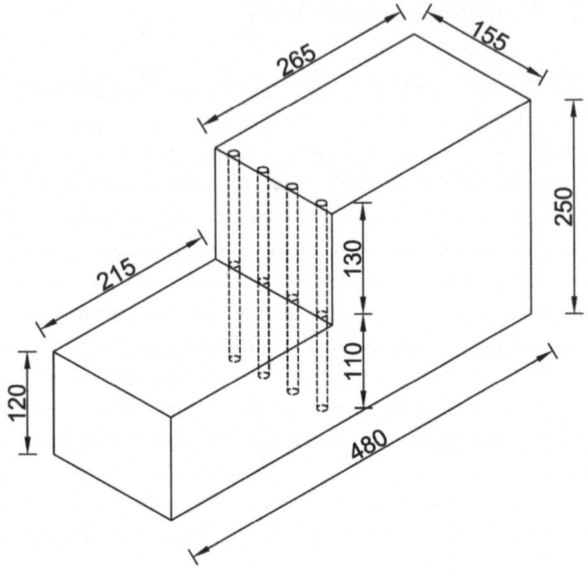

Fig. 1 Geometry of the model (*unit* mm)

applying silicon oil on the contacting surface. The slope shape after failure in Fig. 3b shows that the influence of side friction can be neglected.

Figure 4 presents the total maximum shear strain of the unreinforced slope before failure. The failure started at the toe and propagated to the crest. Even though a nearly plane shear band was formed before the failure, the final failure from the top surface to about a half height of the slope didn't occur along the shear band but along a plane that is almost vertical. The final failure mode for this steep slope is a combination of toppling and sliding.

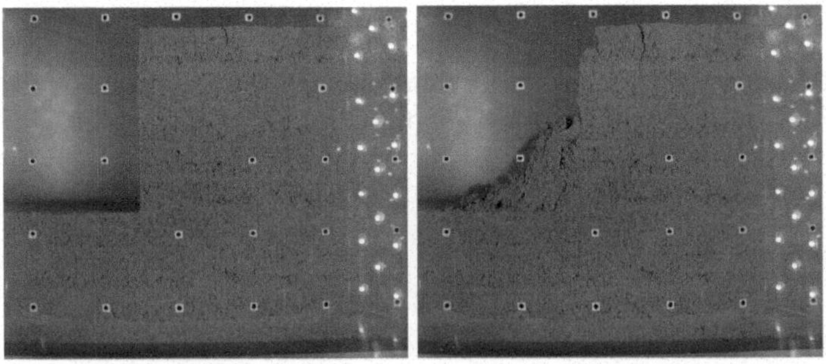

(a) Before the failure (b) After the failure

Fig. 2 The cross sectional plane of the unreinforced slope

3.2 Reinforced Slope

As illustrated in Fig. 5a, the soil behind the piles is divided into three zones. Zone 1 represents the soil that is hardly supported by the piles. The soil in zone 2 is reinforced mainly by the friction between the soil and the pile contacting surface, which corresponds to the frictional soil arching by Li et al. [11]. In zone 3, the main soil arching or end-bearing soil arching by Li et al. [11] is formed. In the centrifuge tests, it was observed that zone 1 failed at a very early stage followed by zone 2 and at last zone 3. Usually the failure of zone 3 represented the failure of the piled slope. Adachi et al. [2] had explained the development of the arching effect in granular soils and pointed out that the uphill pressure does not govern particles B, C, and D (Fig. 5b) within the arching zone of the soil. Li et al. [11] also concluded that the end-bearing soil arching undertakes the most load from the landslide. Numerical studies ([4, 12], etc.) have shown that a big rotation of principle stress direction takes place in zone 3. Thus zone 3 contributes the most to the arching effect and is therefore the area of interest.

(a) Before the failure (b) After the failure

Fig. 3 The top view of the unreinforced slope

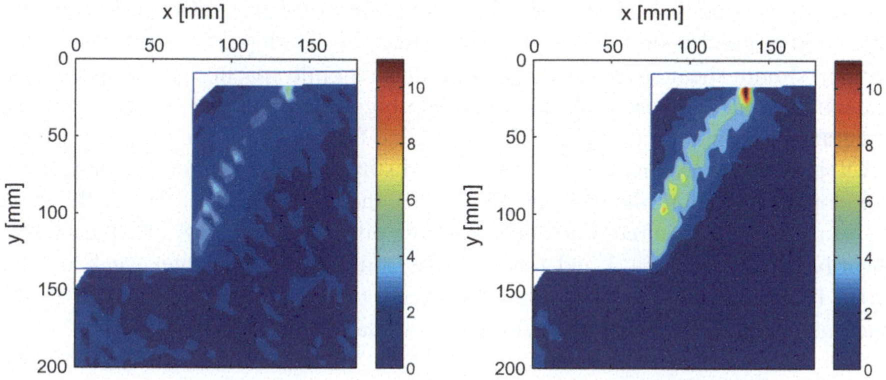

(a) N=7.0 (b) N=7.4

Fig. 4 The total maximum shear strain of the unreinforced slope before failure

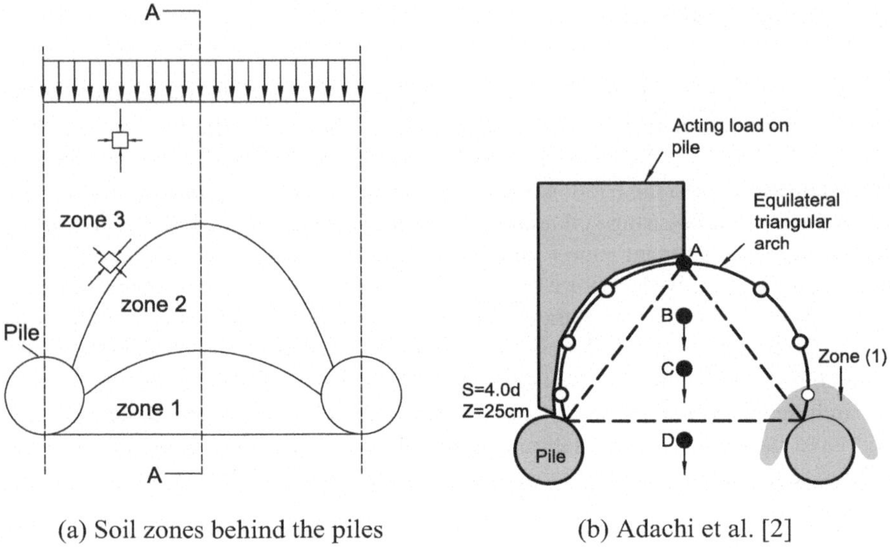

Fig. 5 Arching effect

Figure 6 shows the top view of the failing process of a piled slope (s/d = 3.88). Because of the low pile spacing the friction of the side walls had a big influence on the soil behaviour. As a result, the failure occurred first between the middle piles and the soil close to the back wall and front window followed. As can be seen in Fig. 7, when the pile spacing is big, the boundary of zone 2 is close to or could be even larger than the wedge failure of the slope.

In theory the exposed side view represents the cross section A-A in Fig. 5a, but due to the side friction the failure mass in this plane could be smaller than the cross section in the middle and also later in failing time. However, PIV analysis is still able to give good results of the shearing effect in the slope. Figure 8 shows the total maximum shear strain for slopes with different pile spacing. Multi-shear-band phenomena can be seen in all the tests. The biggest shear band is formed by the integrated deformation of the unstable slope and the others are the local shear band developed by the arching effect, which also distinguish the soil zones along depth. For slopes with small pile spacing (Fig. 8c, d), the shearing effect among the three soil zones is well captured. For slopes with big pile spacing (Fig. 8a, b), the smaller shear band between zone 1 and zone 2 is not formed which indicates weak arching effect from the piles. The shear band between zone 2 and zone 3 is close to the integrated wedge shear band, which are also visualised in Fig. 7.

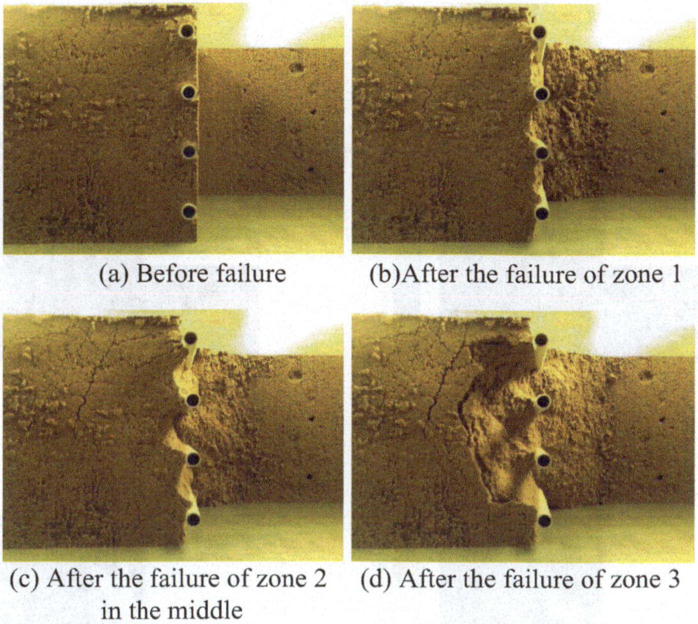

(a) Before failure (b)After the failure of zone 1

(c) After the failure of zone 2 in the middle (d) After the failure of zone 3

Fig. 6 The failure process of a reinforced slope (s/d=3.88)

(a) Failure when s/d = 7.75 (b) Failure when s/d = 15.5

Fig. 7 The failures with big pile spacing

4 Discussion of the Failure of Soil Arching

As pointed out above, the soil in zone 1 and zone 2 contributes less to the soil arching, and therefore is not considered here for simplicity. But it needs to be noted that losing the confinement of zone 1 and zone 2 soil accelerates the failure of zone 3. The soil arching is represented by a half cylindrical shell with a thickness that equals the pile radius, which is a common simplification in piled embankments analyses [1, 6, 13]. Figure 9 shows the sketches of soil arching for different pile spacing. The wedge

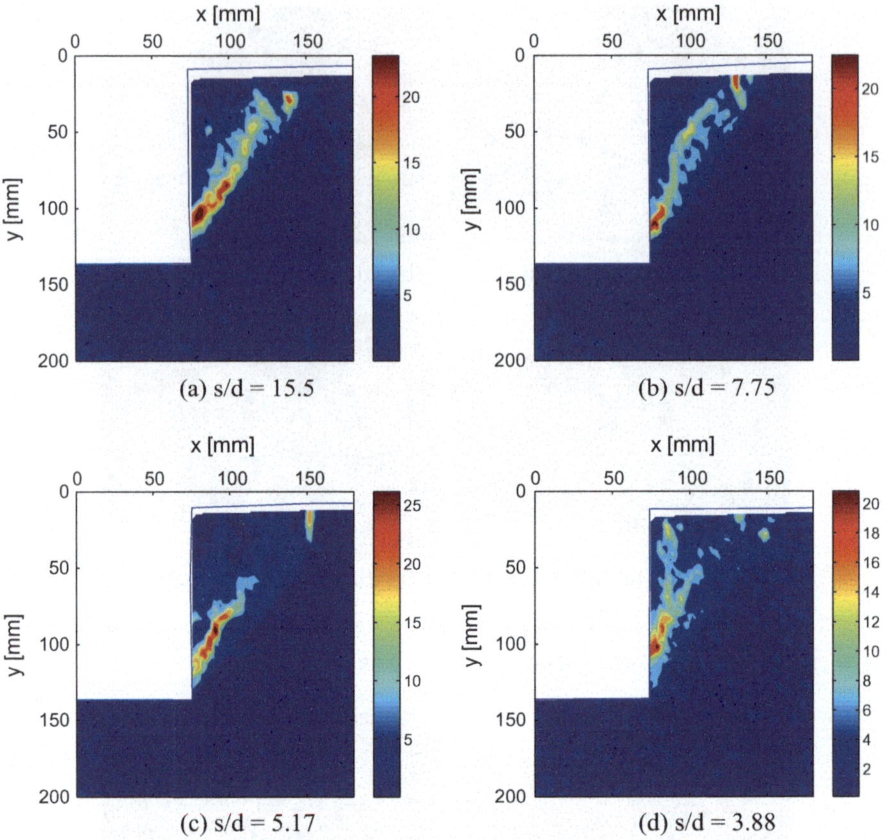

Fig. 8 The total maximum shear strain before failure

failure of the slope is the dominant failure and also the leading factor to the soil arching.

Along the pile axis direction, soil arching is in different status. At the toe of the slope the soil arching is not formed until a certain depth above. At the crest of the slope, the soil arching is fully mobilised. Thus the overall stability of the piled slope depends on the stability of the soil arching near the crest region.

For big pile spacing, the arch boundary is close to the wedge shear band or even outside the shear band. In this case, the dome area is the weak point of the soil arch as can be seen in Fig. 7. Even after the failure, a certain amount of soil is still stabilised by the pile. The piles work similar to single piles. On the contrary, when the spacing is small, the arch boundary is relatively far from the shear band, which gives enough space for load transfer. Under this circumstance the failure will start at the contact between pile and soil and form a smaller wedge failure as in Fig. 6.

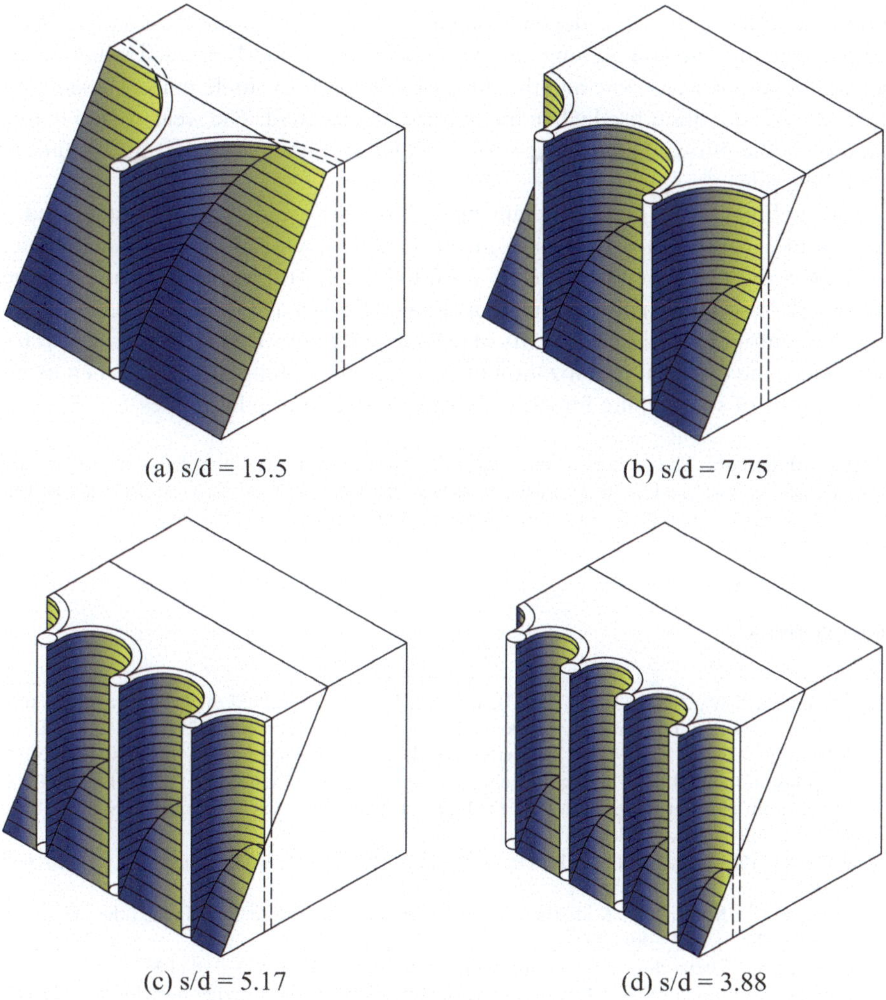

Fig. 9 Sketches of soil arching for different pile spacing

5 Summary and Conclusions

A series of centrifuge tests were conducted to investigate the arching effect in sandy slopes stabilised by a row of discrete piles. The slope went through a plane shearing failure which produced a wedge-shaped moving mass that is divided into three soil zones. Among the three zones, the main soil arching is believed to take place in zone 3. Shearing bands, which distinguish the soil zones and the moving wedge, was found from the PIV analysis of the side view. The wedge shearing is the dominant failure of the slope, which determines that less arching effect in the toe region of the slope and the stability of the slope depends on the arching effect of the crest region. With big pile spacing, the soil arching can not be formed or poorly formed which leads to a failure at the arch dome and the piles work similar to single piles. The arching effect can be fully mobilised when the pile spacing is small. The weak region in this situation is the pile-soil contacting surface which results in a smaller wedge failure behind the piles.

Soil arching is the main mechanism of discrete installed piles to stabilised a slope with excessive horizontal movement. In the view of a 2-D horizontal plane, pile spacing is a control parameter to form the stable soil arching, but in 3-D, the movement of soil is usually not uniform along the depth and the cantilever-working pile also contributes to the complexity of the pile-soil interaction. Thus, it is necessary to consider the different mobilization of arching effect along the depth when using 2-D analysis of soil arching for the stability calculation of piled slopes.

Acknowledgements The authors acknowledge the support from the China Scholarship Council (No.2013 06410004), the EU 7th Framework research project HYDRODRIL (No.295225) as well as the Otto Pregl Foundation for Fundamental Geotechnical Research in Vienna.

References

1. Abusharar, S.W., Zheng, J.-J., Chen, B.-G., Yin, J.-H.: A simplified method for analysis of a piled embankment reinforced with geosynthetics. Geotext. Geomembr. **27**(1), 39–52 (2009)
2. Adachi, T., Kimura, M., TADA, S.: Analysis on the preventive mechanism of landslide stabilizing piles. numerical models in geomechanics. numog iii. Proceedings of the 3rd International Symposium Held in Niagara Falls, Canada, 8–11 May 1989. Elsevier Applied Science Publishers Limited, New York (1989)
3. Bosscher, P.J., Gray, D.H.: Soil arching in sandy slopes. J. Geotech. Eng. **112**(6), 626–645 (1986)
4. Chen, C.-Y., Martin, G.: Soil-structure interaction for landslide stabilizing piles. Comput. Geotech. **29**(5), 363–386 (2002)
5. Handy, R.L.: The arch in soil arching. J. Geotech. Eng. **111**(3), 302–318 (1985)
6. Hewlett, W., Randolph, M.: Analysis of piled embankments. In: International Journal of Rock Mechanics and Mining Sciences and Geomechanics Abstracts, vol. 25, pp. 297–298. Elsevier Science, New York (1988)
7. Ito, T., Matsui, T.: Methods to estimate lateral force acting on stabilizing piles. Soils Found. **15**(4), 43–59 (1975)
8. Kourkoulis, R., Gelagoti, F., Anastasopoulos, I., Gazetas, G.: Slope stabilizing piles and pile-groups: parametric study and design insights. J. Geotech. Geoenviron. Eng. **137**(7), 663–677 (2010)
9. Kourkoulis, R., Gelagoti, F., Anastasopoulos, I., Gazetas, G.: Hybrid method for analysis and design of slope stabilizing piles. J. Geotech. Geoenviron. Eng. **138**(1), 1–14 (2011)
10. Lee, C., Wu, B., Chen, H., Chiang, K.: Tunnel stability and arching effects during tunneling in soft clayey soil. Tunn. Undergr. Space Technol. **21**(2), 119–132 (2006)
11. Li, C., Tang, H., Hu, X., Wang, L.: Numerical modelling study of the load sharing law of anti-sliding piles based on the soil arching effect for Erliban landslide, China. KSCE J. Civil Eng. **17**(6), 1251–1262 (2013)
12. Liang, R., Zeng, S.: Numerical study of soil arching mechanism in drilled shafts for slope stabilization. Soils Found. **42**(2), 83–92 (2002)

13. Low, B., Tang, S., Choa, V.: Arching in piled embankments. J. Geotech. Eng. **120**(11), 1917–1938 (1994)
14. Nadukuru, S.S., Michalowski, R.L.: Arching in distribution of active load on retaining walls. J. Geotech. Geoenviron. Eng. **138**(5), 575–584 (2012)
15. Paik, K., Salgado, R.: Estimation of active earth pressure against rigid retaining walls considering arching effects. Geotechnique **53**(7), 643–654 (2003)
16. Qin, H., Guo, W.D.: Group effects of piles due to lateral soil movement. Int. J. GEOMATE **4**(1), 450–455 (2013)
17. Terzaghi, K.: Theoretical Soil Mechanics. Wiley, New York (1943)
18. Van Eekelen, S., Bezuijen, A., Van Tol, A.: An analytical model for arching in piled embankments. Geotext. Geomembr. **39**, 78–102 (2013)
19. Wang, W.L., Yen, B.C.: Soil arching in slopes. J. Geotech. Eng. **100**(gt1) (1974)
20. White, D., Take, W., Bolton, M.: Soil deformation measurement using particle image velocimetry (PIV) and photogrammetry. Geotechnique **53**(7), 619–632 (2003)

Quantitative Description of Orthotropic Fabric of Sand

Xue-feng Li, Yu-qi He, Jing-qiao Liu and Wei Wu

Abstract Based on the microscopic theory of sand fabric, the novel amplitude parameters describing the spatial arrangement of non-spherical particles is defined, and the verification with the microscopic test for Toyoura sand is conducted. The spatial arrangement of non-spherical sand particles in the orthogonal coordinate is described with two independent angles, one can define three expressions of amplitude parameters which can be quantitatively determined by image analysis on the orthotropic plane. Only two of the three magnitude parameters are independent, with which the three-dimensional orthotropic expression of fabric can be deduced. The expression is used to describe sand properties of orthotropic, transverse isotropic and isotropic. Experimental results show three amplitude parameters with a clear physical meaning can reflect the variation of orthotropic plane.

1 Introduction

The quantitative description of the soil fabric is of fundamental importance. Especially, it is a challenge in the study of orthotropic fabric effects. Available experimental data on soil fabric are scarce. In recent years image analysis has emerged as a promising technique in geotechnical research, because it allows the soil structure to be characterized at a microscopic level. The amplitude parameters research the size of a microscopic fabric orthotropic degree on the surface of metrics. Microscopic fabric is the inner mechanism of the macro mechanical properties,

---

X. Li (✉) · Y. He · J. Liu
Solid Mechanics Institute, Ningxia University, Yinchuan 750021, China
e-mail: lixuefeng1928@163.com

X. Li · W. Wu
Institute of Geotechnical Engineering, University of Natural Resources and Life Sciences, 1180 Vienna, Austria

W. Wu (ed.), *Recent Advances in Geotechnical Research*,
Springer Series in Geomechanics and Geoengineering,
https://doi.org/10.1007/978-3-319-89671-7_10

which has a substantial effect on macroscopic mechanical properties, so it becomes a hot topic [1–10]. A large number of experiments (Guo [1], Pradhan [2], etc.) show that sand microscopic anisotropy obviously control the macroscopic strength characteristics. Wan and Guo [3], Li and Dafalias [4] and a group of scholars all establish the constitutive relationship with fabric tensor. The authors also establish the strength criterion [5], constitutive model [6], property-dependent plastic potential theory for geomaterials [7], and analyze the mechanics characteristics of anisotropic sand [8–10]. The studies of Arthur and Phillips [11], Oda [12], Yamada and Ishihara [13] and Oda and Koishikawa [14] show the random arrangement of non-spherical particles and the distribution of pore space performed by the mathematical laws of probability and statistics, which describe the three causes of sand anisotropy [15]. As for the quantitative test of amplitude parameters, Yang [16] did the experimental work of quantitative determination fabric for Toyoura sand.

Therefore, in order to determine amplitude parameters, based on the microscopic theory of sand fabric, the novel amplitude parameters describing the spatial arrangement of non-spherical particles is defined in this paper.

2 The Definition of Granular Material Fabric Tensor

Random arrangement of non-spherical particles and its void ratio have the fabric properties of the mathematical statistical significance. In order to facilitate the related characteristics of fabric, Oda [14, 15] and Tobita [17] defined the fabric tensor of the F_{ij} with the effect of non-spherical particles arrangement:

$$F_{ij} = \int_{\Omega} n_i n_j E(\vec{n}) d\Omega \; (i, j = 1, 2, 3), \tag{1}$$

where $E(\vec{n})$ is the probability density function of the unit vector $\vec{n}$, and $\vec{n}$ is the component of non-spherical particles on long axis direction. F_{ij} is symmetric, i.e. $F_{ij} = F_{ji}$ and $F_{ii} \equiv 1$.

By the relation of integral and series, the integral expression can be had equivalent calculation by series superposition, Eq. (1) can be rewritten as:

$$F_{ij} = \frac{1}{2N} \sum_{k=1}^{2N} n_i^{(k)} n_j^{(k)} \; (i, j = 1, 2, 3) \tag{2}$$

where N is the number of particles in the measurement volume; $n_i^{(k)}$ and $n_i^{(k)}$ represent the component of normal vector of particles long axis on i and j axis, respectively. The three angles $\theta_1^{(k)}$, $\theta_2^{(k)}$ and $\theta_3^{(k)}$ as shown in Fig. 1a, but only two angles of three

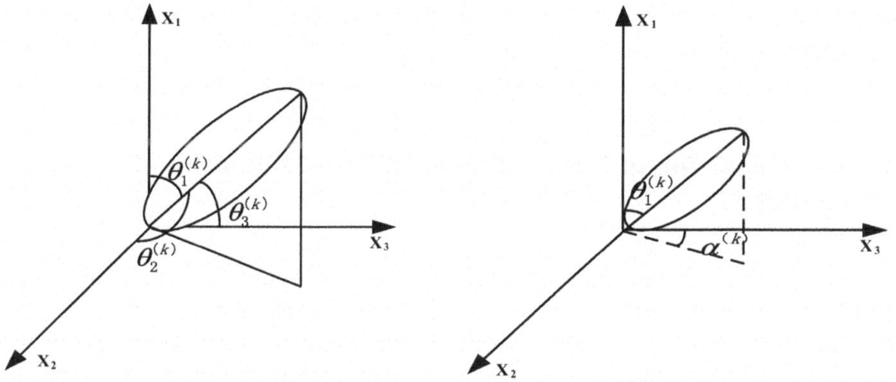

Fig. 1 The orientation components of particles

are independent variables. Therefore, the expressions of three-dimensional fabric can be calculated only with two angles $\theta_1^{(k)}$ and $\alpha^{(k)}$ [18]. As shown in Fig. 1b, $\alpha^{(k)}$ is the angle between the long axis of kth particle projection in the horizontal plane and the X_3 axis. So the three angles representation of the fabric tensor can be turned into a two angles representation. The direction vector of F_{ij} is:

$$\boldsymbol{n} = \left\{\cos\theta_1^{(k)}, \sin\theta_1^{(k)}\sin\alpha^{(k)}, \sin\theta_1^{(k)}\cos\alpha^{(k)}\right\} \tag{3}$$

The component of fabric in Eq. (2) can be expressed as:

$$F_{ij} = \begin{bmatrix} F_{11} & F_{12} & F_{13} \\ F_{21} & F_{22} & F_{23} \\ F_{31} & F_{32} & F_{33} \end{bmatrix}$$
$$= \begin{bmatrix} \frac{1}{2N}\sum_{k=1}^{2N}\cos^2\theta_1^{(k)} & \frac{1}{4N}\sum_{k=1}^{2N}\sin 2\theta_1^{(k)}\sin\alpha^{(k)} & \frac{1}{4N}\sum_{k=1}^{2N}\sin 2\theta_1^{(k)}\cos\alpha^{(k)} \\ \frac{1}{4N}\sum_{k=1}^{2N}\sin 2\theta_1^{(k)}\sin\alpha^{(k)} & \frac{1}{2N}\sum_{k=1}^{2N}\sin^2\theta_1^{(k)}\sin^2\alpha^{(k)} & \frac{1}{4N}\sum_{k=1}^{2N}\sin^2\theta_1^{(k)}\sin 2\alpha^{(k)} \\ \frac{1}{4N}\sum_{k=1}^{2N}\sin 2\theta_1^{(k)}\cos\alpha^{(k)} & \frac{1}{4N}\sum_{k=1}^{2N}\sin^2\theta_1^{(k)}\sin 2\alpha^{(k)} & \frac{1}{2N}\sum_{k=1}^{2N}\sin^2\theta_1^{(k)}\cos^2\alpha^{(k)} \end{bmatrix} \tag{4}$$

Equation (4) extended the component spatial of single particle to general space based on the fabric theory definition, which described the mathematical law of statistics of particles arrangement and the orthogonal anisotropic produced by the arrangement characteristics of particles theoretically.

3 The Definition of Fabric Tensor in Two-Dimensional Spatial

For orthotropic anisotropic materials, three orthogonal planes should have amplitude parameters in each plane. As shown in Fig. 2, the three orthogonal planes $F_{x1} - F_{x2}$, $F_{x1} - F_{x3}$ and $F_{x2} - F_{x3}$ have an anisotropic amplitude parameter respectively.

According to defined method of amplitude parameters in two-dimensional spatial, by Eq. (4), we can obtain:

$$\overline{F}_{1,3} = \frac{1}{2}(F_{11} + F_{33}) \pm \left[\frac{1}{4}(F_{11} - F_{33})^2 + F_{13}^2\right]^{\frac{1}{2}} = \frac{1 \pm a_1}{2} \tag{5}$$

where

$$a_1 = \frac{1}{2N}\sqrt{\left(\sum_{k=1}^{2N}\left(\cos^2\theta_1^{(k)} - \sin^2\theta_1^{(k)}\cos^2\alpha^{(k)}\right)\right)^2 + \left(\sum_{k=1}^{2N}\sin 2\theta_1^{(k)}\cos\alpha^{(k)}\right)^2} \tag{6}$$

with the same idea, we can define the other two parameters of two-dimensional spatial in the planes of $F_{x1} - F_{x2}$ and $F_{x2} - F_{x3}$.

$$a_2 = \frac{1}{2N}\sqrt{\left(\sum_{k=1}^{2N}\left(\cos^2\theta_1^{(k)} - \sin^2\theta_1^{(k)}\sin^2\alpha^{(k)}\right)\right)^2 + \left(\sum_{k=1}^{2N}\sin 2\theta_1^{(k)}\sin\alpha^{(k)}\right)^2} \tag{7}$$

$$a_3 = \frac{1}{2N}\sqrt{\left(\sum_{k=1}^{2N}\left(\sin^2\theta_1^{(k)}\cos^2 2\alpha^{(k)}\right)\right)^2 + \left(\sum_{k=1}^{2N}\sin^2\theta_1^{(k)}\cos 2\alpha^{(k)}\right)^2} \tag{8}$$

where a_1, a_2 and a_3 are the anisotropic amplitude parameters in three orthogonal planes, respectively. The three parameters can be determined by Eqs. (6), (7) and (8) with microscopic test, the three parameter value range is 0–1. Three anisotropic amplitude parameters describe the projection component of the long axis for all the non-spherical particles, with the ideas of the mathematical probability statistics calculation, which depict the direction of irregular discrete particle in the plane. Its physical meaning is the spatial arrangement of random particles with mathematical statistical sense of the orthotropic; the anisotropic value parameter is the microscopic characteristic of discrete particles aggregate macro performance on the plane. This value is a scalar, and can be directly applied to macroscopic mechanical equations.

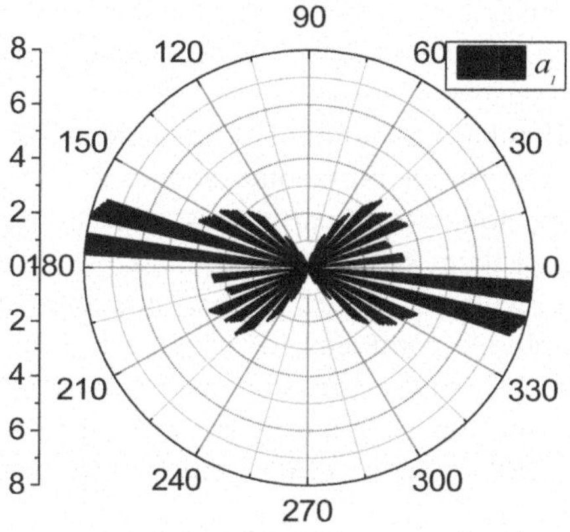

(a) Result of a_1

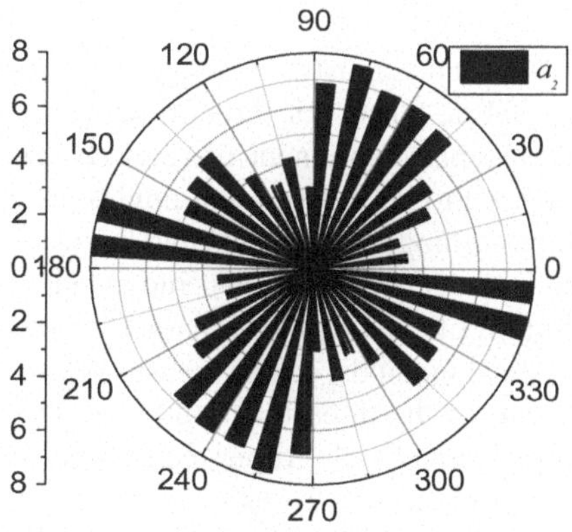

(b) Result of a_2

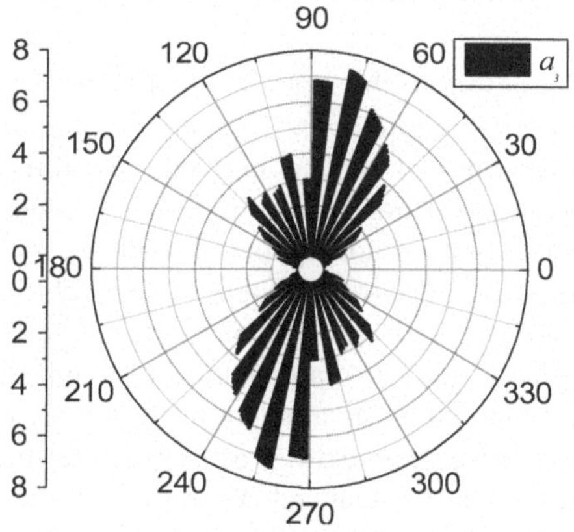

(c) Result of a_3

Fig. 2 Rose diagram of amplitude parameters with $\alpha^{(k)} = \pi/2$ for dry deposited specimen

4 Characteristic Analysis of Three-Dimensional Fabric Tensor

We select the results of Yang [17] who engaged in sand fabric with a quantitative analysis. In this study Toyoura sand, a Japanese standard sand consisting of sub-rounded to subangular particles was used. Both the dry deposition (DD) and moist tamping (MT) methods were employed to produce specimens for laboratory testing by scanning electron microscopy (SEM) for horizontal and vertical direction sample pictures. Through image processing, each particle was numbered and its orientation was identified. By a statistical analysis of a number of images of the vertical sections for both DD and MT coupons, the values of the anisotropic amplitude parameters were obtained.

Figure 2 shows the simulation of amplitude parameter which describe the preferred orientation of the particles for the horizontal sections of DD coupons. In this paper, the coordinate axes of the long axis of particles is different from the axis of reference axes. The relationship of them satisfies $\theta^{(k)} = \pi/2 - \theta'^{(k)}$. The Figure shows the calculation of a_1, a_2, a_3 with $\theta^{(k)} = \pi/2$. In this situation, the parameters of a_1 and a_2 are the vertical amplitude parameter, of which the form is consistent with the definitions of Oda [14, 15], so their simulation results are also identical. Figure 2 shows that the simulation of a_1, a_2 and a_3 are vertical between them, which also indirectly validated the orthogonality of the defined three parameters in this article.

5 Conclusion

According to the mathematical laws of probability and statistics belonged to non-spherical particles in the spatial, a quantitative fabric tensor for three-dimensional assemblies of sand is derived. The three-dimensional component of referred orientation for non-spherical particles is described with two independent angles. The component of non-spherical particles in the three orthogonal surfaces can be quantified by anisotropy magnitude of fabric which is deduced with two independent angles in each orthogonal surface. The description of sand discrete characteristics, with the discrete mathematical probability statistical methods, can be measured with the magnitude parameters in the plane. Only two of the three magnitude parameters are independent, with which the three-dimensional orthotropic expression of fabric can be deduced. With the change in the physical properties of microscopic particles, three-dimensional expression of the fabric can describe isotropic material, transversely isotropic and orthotropic naturally. The microscopic test methods of the quantitative magnitude parameters is proposed in the paper. Analyses showed that the proposed fabric has a wide range of characteristics and a clear physical meaning.

Acknowledgements This work was financially supported by the National Key R&D Program of China (2017YFC0504400, 2017YFC0504404) and the National Natural Science Foundation of China (5168050, 51669027). These supports are gratefully acknowledged.

References

1. Guo, P.J.: Modified direct shear test for anisotropic strength of sand. J Geotech Geoenviron Eng, ASCE **134**(9), 1311–1318 (2008)
2. Pradhan, T.B.S., Tatsuoka, F., Horii, N.: Strength and deformation characteristics of sand in torsion simple shear. Soils Found. **28**(3), 131–148 (1985)

3. Wan, R.G., Gua, P.J.: A pressure and density dependent dilatancy model for granular materials. Soils Found. **39**(6), 1–12 (1999)
4. Li, X.S., Dafalias, Y.F.: Constitutive modeling of inherently anisotropic sand behavior. J. Geotech. Geoenviron. Eng., ASCE **128**(10), 868–880 (2002)
5. Li, X.F., Huang, M.S., Qian, J.G.: Failure criterion of anisotropic sand with the method of macro-micro incorporation. Chin. J. Rock Mechan. Eng. **29**(9), 1885–1892 (2010)
6. Huang, M.S., Li, X.F., Qian, J.G.: On strain localization of anisotropic sands. Chin. J. Geotech. Eng. **34**(10), 1885–1892 (2012)
7. Li, X.F., Kong, L., Huang, M.S.: Property-dependent plastic potential theory for geomaterials. Chin. J. Geotech. Eng. **35**(9), 1722–1729 (2013)
8. Li, X.F., Huang, M.S., Qian, J.G.: Analysis of non-coaxial characters of sand for simple shear test with the method of macro-meson-incorporation. Rock Soil Mech. **34**(12), 3417–3424 (2013)
9. Li, X.F., Huang, M.S., Kong, L.: Failure properties of sand considering rotation of principal stress axis with method of macro-meso incorporation. Rock Soil Mech. **34**(7), 1923–1930 (2013)
10. Li, X.F., Huang, M.S., Qian, J.G.: Strain localization analysis of anisotropic sands based on non-coaxial theory. Eng. Mech. **31**(3), 205–211 (2014)
11. Arthur, J.R.F., Phillips, A.B.: Correspondence: inherent anisotropy in sand. Geotechnique **22**(3), 537–539 (1972)
12. Oda, M.: Initial fabrics and their relations to mechanical properties of granular materials. Soils Found. **12**(1), 17–36 (1972)
13. Yamada, Y., Ishihara, K.: Anisotropic deformation characteristics of sand under three-dimensional stress conditions. Soils Found. **19**(2), 79–94 (1979)
14. Oda, M., Koishikawa, I.: Anisotropic fabric of sands. Soils Found. **17**(4), 71–77 (1977)
15. Oda, M.: Fabric tensor and its geometrical meaning. In: Oda, M., Iwashitak, K. (eds.) Introduction to mechanics of granular materials, pp. 27–35. A. A. Balkema, The Netherlands (1999)
16. Yang, Z.X., Li, X.S., Yang, J.: Quantifying and modeling fabric anisotropy of granular soils. Geotechnique **58**(4), 237–248 (2008)
17. Tobita, Y.: Contact tensor in constitutive model for granular materials. In: Satake, M., Jenkins, J. (eds.) Proceedings U.S.-Japan seminar on micromechanics of granular materials, Sendai-Zao, Japan, Elsevier, New York, pp. 263–270 (1988).
18. Li, X.F., Wang, X., Yuan, Q.: Quantitative determination of rock crack fabric. Chin. J. Rock Mechan. Eng. **34**, 2355–2361 (2015)

A Novel Description of Plastic Strain Direction

Xue-feng Li, Yu-qi He, Liang Kong, Wei Wu and Yan-chang Wang

Abstract The paper presents a plastic flow equation based on a property-dependent potential theory for geomaterials to describe the flow of sand during loadings involving rotation of principal stress direction. To overcome the shortage of the traditional potential theory with the assumption of isotropy, the property-dependent plastic potential theory is proposed by linking the strain distribution law with the material properties described by the fabric tensor. Based on the proposed potential theory, the plastic flow equation is derived from the energy dissipation in the state of the critical state, which is the function of loading stress, the degree of fabric anisotropy and the geometric relationship between fabric and stress. Therefore the proposed plastic flow equation can not only describe the uniqueness flow when the fabric is isotropic, but can also describe the non-uniqueness flow or the dependency of the plastic strain increment direction on the stress increment direction when the fabric is anisotropic. Compared with the existing non-coaxial plasticity theory, the proposed potential theory has a clearer physical meaning.

X. Li (✉) · Y. He
Solid Mechanics Institute, Ningxia University, Yinchuan 750021, China
e-mail: lixuefeng1928@163.com

X. Li · W. Wu
Institute of Geotechnical Engineering, University of Natural Resources and Life Sciences, 1180 Vienna, Austria

L. Kong
School of Sciences, Qingdao Technological University, Qingdao 266033, China

Y. Wang
School of Physics and Electronic-Electrical Engineering, Ningxia University, Yinchuan 750021, Ningxia, China

W. Wu (ed.), *Recent Advances in Geotechnical Research*,
Springer Series in Geomechanics and Geoengineering,
https://doi.org/10.1007/978-3-319-89671-7_11

1 Introduction

The principal plastic directions will seldom flow along the principal stress direction, which is a common aspect of many loading situations in cyclic loading environments caused by earthquakes, vehicular traffic and sea waves. The traditional elastoplastic theory fails to describe such plastic flow attribute to its potential theory. As the direction hypothesis of plastic strain, the traditional potential theory ensures the coincidence of the direction of plastic flow with the plastic potential function that meets the associated flow rule. However, for such assumptions of traditional potential theory, there are lots of defects when it is applied to the constitutive model. Therefore, we try to use a novel potential theory to overcome the defects of traditional potential theory.

A large amount of soil experiments show that traditional plastic potential theory fails to describe the deformation behavior caused by the assumptions, such as continuity, homogeneity and isotropy [1]. The continuity assumptions lead to the shortage of traditional potential theory, through which is difficult to describe discrete-related features of soil. In order to apply the microstructure theory on macro phenomenon, a large number of scholars are committed to DEM in the recent years [2, 3]. The isotropy hypothesis fails to describe the anisotropic and its related features. Whether soil particles have a uniform distribution or not, there are two kinds of mechanical properties, isotropy and anisotropy, and anisotropy makes the powerful influence. Because of the anisotropy of soil, the rotation of the principal stress axes is generated and non-coaxial features are showed at the same time. The plastic potential theory based on isotropic hypothesis cannot calculate the plastic deformation produced by the rotation of the principal stress axes, and it also fails to describe the non-coaxial features. In order to solve these problems, a lot of scholars are committed to the research. For example, the generalized potential theory proposed by Yang [4], the generalized plastic mechanics considering the rotation of the principal stress axes established by Zheng [1, 5], the non-coaxial theory proposed by Rudnick [6], the non-coaxial model proposed by Yang [7] and Huang [8]. Gutierrez [9] and Shi [10] introduced the novel non-coaxial plastic theory [9]. All these ideas have the shortages to different degrees, such as the complex forms of mathematics, the parameters without a clearer physical meaning, and so on. As for the mechanism of non-coaxial, the reason of which is the existence of anisotropy, and the anisotropy of soil can be attributed to the micromechanical properties. From the views of all the above, such as the rotation of principal stress axes and non-coaxial correction, are based on the isotropic hypothesis of continuum mechanics to describe. Therefore how to deal with isotropic hypothesis is crucial to describe the anisotropy and its related characteristics.

Aiming at above problems of the traditional plastic potential theory, based on property-dependent potential theory for geomaterials and the research on mechanical properties of sand [11–14], we established the plastic flow equation, derived from energy dissipation with the state of critical state, and analyzed the direction uniqueness of the strain increment.

2 Property-Dependent Plastic Potential Theory for Geomaterials

2.1 Physical Property

The fabric of geomaterials exist objectively. The fabric tensor can be obtained by test in the three principal directions, so we [14] obtained the fabric expression

$$F_{ij} = \frac{1}{3 - a_1 - a_3 - a_1 a_3}\begin{bmatrix} 1 + a_1 - a_3 - a_1 a_3 & 0 & 0 \\ 0 & 1 - a_1 + a_3 - a_1 a_3 & 0 \\ 0 & 0 & 1 - a_1 - a_3 + a_1 a_3 \end{bmatrix} \quad (1)$$

where a_1 and a_3 are the amplitude parameters of anisotropy. Based on the mathematical statistical laws, a_1 and a_3 are projections component of the individual particles on the plane. Individual particles can be described by the independent angles of $\theta_1^{(k)}$ and $\alpha^{(k)}$.

$$a_1 = \frac{1}{2N}\sqrt{\left(\sum_{k=1}^{2N}\left(\cos^2\theta_1^{(k)} - \sin^2\right)\right)^2 + \left(\sum_{k=1}^{2N}\sin 2\theta_1^{(k)}\cos\alpha^{(k)}\right)^2} \quad (2)$$

$$a_3 = \frac{1}{2N}\sqrt{\left(\sum_{k=1}^{2N}\left(\sin^2\theta_1^{(k)}\cos^2 2\alpha^{(k)}\right)\right)^2 + \left(\sum_{k=1}^{2N}\sin^2\theta_1^{(k)}\cos 2\alpha^{(k)}\right)^2} \quad (3)$$

The amplitude parameters can be determined by microscopic experiments from three orthogonal planes respectively. The value of three parameters are 0 to 1 to describe the degree of anisotropy, which is determined by from Eqs. (2) and (3). For the orthotropic anisotropic materials, there are three amplitude parameters on the orthogonal planes.

2.2 The Property-Dependent Plastic Potential Theory

The main idea of property-dependent plastic potential theory is based on the presumption that the material properties can be described by using a fabric tensor, so the rules of strain distribution will be affected by the material properties. Therefore the expression of property-dependent plastic potential theory for geomaterials [13] can be expressed as

$$d\varepsilon_{ij}^p = d\lambda\frac{\partial g}{\partial\sigma_{ij}}\overline{F}_{ij} \quad (4)$$

where g is the plastic potential function, and $d\lambda$ is the factor of incremental plastic. $\overline{F}_{ij}$ is the direction tensor of fabric tensor F_{ij} in Eq. (1).

According to the characteristics of $\overline{F}_{ij}$, if the material is isotropic, $\overline{F}_{ij} \equiv \delta_{ij}$, δ_{ij} is the Kronecker tensor (when $i \neq j$, $\delta_{ij} = 0$; when $i = j$, $\delta_{ij} = 1$). The Eq. (4) degenerates to the form of traditional potential theory with the isotropic hypothesis. When the material is anisotropic, the law of plastic strain distribution in Eq. (4) is affected by both the degree and the direction of material anisotropy.

According to the characteristics of tensor, the fabric F_{ij} in Eq. (1) can be expressed as

$$F_{ij} = p_F(\delta_{ij} + \tilde{s}_{ij}) = p_F(\overline{F}_{ij}) \tag{5}$$

Where $\overline{F}_{ij}$ and p_F are the direction and the degree of the fabric tensor respectively, $p_F{=}1/F_{kk}$, $\overline{F}_{ij} = \delta_{ij} + \tilde{s}_{ij}$. Introducing the fabric tensor $\overline{F}_{ij}$ of Eq. (5) into Eq. (4), we can get another form of the potential theory as

$$d\varepsilon_{ij}^{p} = d\lambda\frac{\partial g}{\partial \sigma_{ij}} + d\lambda\frac{\partial g}{\partial \sigma_{il}}\tilde{s}_{lj} \tag{6}$$

Equation (6) shows that the plastic strain increment can be decomposed into two parts, namely, isotropy and anisotropy. If the material is anisotropic, Eq. (6) describe the effect both the degree and the direction of fabric on the distribution of plastic strain increment. From Eq. (6), the plastic dissipated work is naturally divided into two parts, the isotropic plastic work and the plastic dissipation work related to the material properties if they are calculated by the proposed potential theory. Therefore, the calculation component of plastic dissipation work by the plastic potential theory is the function of the stress component and the direction between the stress and the direction of plastic strain increment. The calculation component is different for the changes of plastic dissipation function under the condition of principal stress axis rotation.

3 The Plastic Potential Function of Property-Dependent Plastic Potential Theory

Conservation of energy is reasonable to explain the deformation of rock and soil under any conditions as the most basic laws of nature. Assuming geotechnical materials is loaded under any stress σ_{ij}, the plastic dissipation energy dW^p can be expressed [1, 13]:

$$dW^p = \sigma_{ij} d\varepsilon_{ij}^{p} \tag{7}$$

Introducing Eq. (6) into Eq. (7), we can obtain the expression for plastic dissipation work related to material properties:

$$dW^p = \sigma_{ij} d\varepsilon_{ij}^p = \sigma_{ij}(d\lambda \frac{\partial g}{\partial \sigma_{ij}} + d\lambda \frac{\partial g}{\partial \sigma_{il}} \tilde{s}_{lj}) \tag{8}$$

With the express of principal stress, Eq. (8) can also be written in the following form

$$d\overline{W}^p = \sigma_k d\varepsilon_k^p = \sigma_k(d\lambda \frac{\partial g}{\partial \sigma_k} + d\lambda \frac{\partial g}{\partial \sigma_k} \tilde{s}_k) \tag{9}$$

Although Eq. (9) still keeps the plastic strain increment and stress in the same direction, however, it can describe the non-coaxial characteristics of materials with the changes of fabric component. To facilitate understanding, we converted Eq. (9) into the form of stress invariants (p, q, θ_σ) and the corresponding strain increment invariants $\left(\varepsilon_v^p, \gamma^p, \theta_{d\varepsilon}\right)$.

The principal stress can be expressed by its invariant as

$$\sigma_k = p + \frac{2}{3} q \sin\left[\theta_\sigma + \frac{2\pi(2-k)}{3}\right]; k = 1, 2, 3 \tag{10}$$

where $p = \sigma_{kk}/3$, $q = \sqrt{3J_2}$, $J_2 = s_{ii}s_{ii}/2$, $s_{ij} = \sigma_{ij} - \delta_{ij} p$, $\theta_\sigma = \sin^{-1}(3\sqrt{3}J_3/2J_2^3)/3$.

Also, the plastic strain increment can be expressed by its invariant as

$$d\varepsilon_k^p = \frac{1}{3} d\varepsilon_v^p + d\gamma^p \sin\left[\theta_{d\varepsilon} + \frac{2(2-k)\pi}{3}\right] \tag{11}$$

where $\mathrm{d}\varepsilon_v^{\mathrm{p}} = \mathrm{d}\varepsilon_{kk}^{\mathrm{p}}/3$, $\mathrm{d}\gamma^{\mathrm{p}} = \sqrt{3J_2'/4}$, $J_2' = \mathrm{d}e_{ij}^{\mathrm{p}}\mathrm{d}e_{ij}^{\mathrm{p}}/2$, $\mathrm{d}e_{ij}^{\mathrm{p}} = \mathrm{d}\varepsilon_{ij}^{\mathrm{p}} - \delta_{ij}\mathrm{d}\varepsilon_{ij}^{\mathrm{p}}/3$, $J_3' = \mathrm{d}e_{ij}^{\mathrm{p}}\mathrm{d}e_{ik}^{\mathrm{p}}\mathrm{d}e_{ki}^{\mathrm{p}}$, $\theta_{d\varepsilon} = \sin^{-1}(3\sqrt{3}J_3'/2J_2'^{3/2})/3$.

Introducing Eqs. (10) and (11) into Eq. (9), the following equation can be obtained:

$$\begin{aligned} \mathrm{d}\overline{W}^{\mathrm{p}} &= p\mathrm{d}\varepsilon_v^{\mathrm{p}} + \frac{2q\mathrm{d}\gamma^{\mathrm{p}}}{3}\sum_{k=1}^{3} \sin\left[\theta_\sigma + \frac{2(2-k)\pi}{3}\right] \sin\left[\theta_{\mathrm{d}\varepsilon} + \frac{2(2-k)\pi}{3}\right] \\ &= p\mathrm{d}\varepsilon_v^{\mathrm{p}} + q\mathrm{d}\gamma^{\mathrm{p}} \cos(\theta_\sigma - \theta_{\mathrm{d}\varepsilon}) \end{aligned} \tag{12}$$

4 Analysis of the Plastic Strain Direction

Figure 1 shows the plastic strain direction of proposed potential theory. According to Eq. (4), the component of plastic strain increments are determined by the plastic factor $\mathrm{d}\lambda$, fabric tensor $\overline{F}_k$ and stress gradient direction $\partial g/\partial \sigma_k$, $k = 1, 2, 3$. When

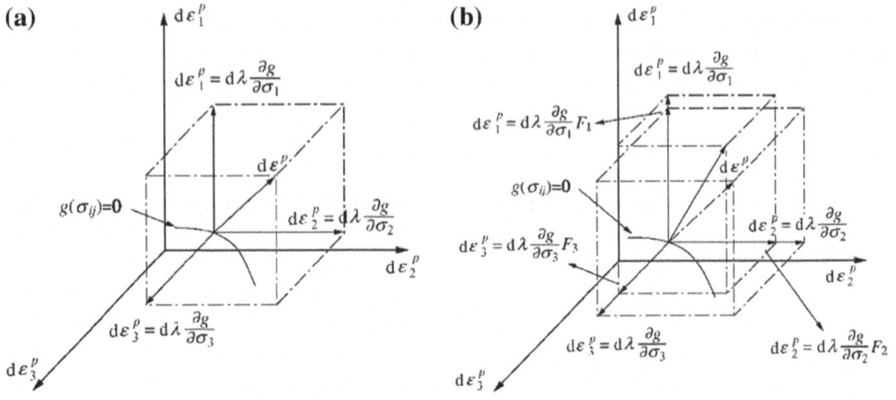

Fig. 1 **a** The direction of traditional potential theory; **b** The direction of property-dependent potential theory

the material is isotropic, $\overline{F}_{ij} \equiv \delta_{ij}$ then there is no effect of fabric on total direction of plastic strain but only on the stress state. When the material is anisotropic, although the component direction of plastic strain is the same in stress gradient direction, the anisotropy size leads to a change of total direction of plastic strain. Therefore, the total plastic strain direction is determined by stress gradient and material properties, and the strain increment direction changes with the degree of material anisotropy and stress direction.

5 Conclusion

Aiming at the shortages of traditional potential theory caused by the assumptions for geomaterials, in the framework of continuum mechanics theory, we establish the property-dependent plastic potential theory by linking the strain distribution with the material properties. The material properties are described by the fabric tensor from the microscopic. When the material is isotropic, the property-dependent plastic potential theory degrades to the traditional potential theory. When the material is anisotropic, the plastic strain increment is the function of loading stress, the degree of fabric anisotropy and the geometric relationship between fabric and stress. Therefore, the proposed theory can not only describe anisotropic distribution of the plastic strain increment under principal stress fixed conditions, but can also describe the values and direction of the plastic strain increment under principal stress rotation conditions, and it has a clearer physical meaning.

Acknowledgements This work was financially supported by the National Key R&D Program of China (2017YFC0504400, 2017YFC0504404) and the National Natural Science Foundation of China (5168050, 51669027). These supports are gratefully acknowledged.

References

1. Zheng, Y., Kong, L.: Geotechnical plastic mechanics. China Architecture and Building Press, Beijing (2010)
2. Cheng, Y.P., Nakata, Y., Bolton, M.D.: Discrete element simulation of crushable soil. Geotechnique **53**, 633–641 (2003)
3. Jiang, M.J., Yu, H.-S., Harris, D.: A novel discrete model for granular material incorporating rolling resistance. Comput. Geotech. **32**, 340–357 (2005)
4. Yang, G., Li, G., Jie, Y.: Soil constitutive model of generalized potential theory and its application. China Water Resources and Hydropower Press, Beijing (2007)
5. Liu, Y., Zheng, Y.: Generalized plastic potential theory involving the rotation of principal stress axes. Chin. Q. Mech. **21**, 129–133 (2001)
6. Rudnicki, J.W., Rice, J.R.: Conditions for localization of deformation in pressure-sensitive dilatant materials. J. Mech. Phys. Solids **23**, 371–394 (1975)
7. Yang, Y., Yu, H.S.: A non-coaxial critical state soil model and its application to simple shear simulation. Int. J. Numer. Anal. Meth. Geomech. **30**, 1369–1390 (2006)
8. Huang, Maosong, Lu, Xilin, Qian, Jiangu: Non-coaxial elasto-plasticity model and bifurcation prediction of shear banding in sands. Int. J. Numer. Anal. Methods Geomech. **34**, 906–919 (2010)
9. Gutierrez, M., Ishihara, K.: Non-coaxiality and energy dissipation in granular materials. Soils Found. **40**, 49–59 (2000)
10. Shi, H., Xie, D., Wang, W.: Strain due to rotation of principal stress axes under plane strain condition. Chin. J. Geotech. Eng. **23**, 162–166 (2001)
11. Li, X., Huang, M., Qian, J.: Failure criterion of anisotropic sand with the method of macro-micro incorporation. Chin. J. Rock Mechan. Eng. **29**, 1885–1892 (2010)
12. Li X., Huang M., Kong L.: Sand failure properties under principal stress rotation condition with the method of macro-micro incorporation. Rock Soil Mech. (2012)
13. Li, X., Kong, L., Huang, M.: Property-dependent plastic potential theory for geomaterials. Chinese. J. Geotech. Eng. **35**, 1722–1729 (2013)
14. Li, X., Wang, X., Yuan, Q.: Quantitative determination of rock crack fabric. Chin. J. Rock Mechan. Eng. **34**, 2355–2361 (2015)

Experimental and Numerical Study on Heat Transfer of Soil Around Underground Cable

Jia Lin, Andreas Bolzer, Herbert Schort, Markus Rauchecker and Wei Wu

Abstract In Europe during summer time, there are very often outages of underground cable. Due to the high energy consumption in summer, the cable will generate great amount of heat to the surrounding soils. The heat lead to drying of the soils and decrease the thermal conductivity of soil. Overheating of the cable will cause outages of the power network. The interaction of the underground cable and soil is very complex. The aim of this study is to use experimental and numerical methods to model the heat transfer problems of the underground cable. The thermal conductivities of soils with different water contents are measured. The numerical models are calibrated by the model tests.

1 Introduction

Underground cable is the replacement of overhead cables providing electrical power or telecommunications. This is typically performed for aesthetic purposes, but also

J. Lin (✉) · M. Rauchecker · W. Wu
Institute of Geotechnical Engineering, University of Natural Resources and Life Sciences, Feistmantelstrasse 4, 1180 Vienna, Austria
e-mail: jia.lin@boku.ac.at

M. Rauchecker
e-mail: markus.rauchecker@boku.ac.at

W. Wu
e-mail: wei.wu@boku.ac.at

A. Bolzer · H. Schort
Wiener Netze GmbH, Wien, Austria
e-mail: andreas.bolzer@wienernetze.at

H. Schort
e-mail: herbert.schort@wienernetze.at

W. Wu (ed.), *Recent Advances in Geotechnical Research*,
Springer Series in Geomechanics and Geoengineering,
https://doi.org/10.1007/978-3-319-89671-7_12

serves the additional significant purpose of making the power lines less susceptible to outages during high wind thunderstorms or heavy snow or ice storms. In Austria, low and medium- voltage lines are most secure underground, while many believe it makes sense to leave high voltage wires high above ground. Till the end of year 2011, the total length of the medium voltage (up to 110 kV) power lines in the Austrian public electricity grid was about 72871 km, out of which about 67% lines are underground. Almost all the distribution lines in Austria are already undergrounded.

During the power transmission of underground cables, great amount of heat is generated in the cables, according to Joule's law. If the heat cannot be dissipated by the surrounding soil, the cable will burn out. This normally happens in summer time, when the soil is hot and the cables are overloaded. Since almost all transmission and distribution lines within the Vienna city area are underground, there is a frequent power outage due to overheated power cables. A study carried out by Reichl et al. [1] showed that the total cost of a 10 h power outage in Austria will be about 502 Million Euro and the value of lost load per hour in summer will be €17.1/ kWh. In order to prevent this, the heat transfer procedures between underground cable and soils need to be studied.

The heat conduction is described as

$$\rho C \frac{\partial T}{\partial t} - \nabla(\lambda \nabla T) = Q \tag{1}$$

where T is the temperature, t is the time, ρ is the density, C is the specific heat, $\nabla = \frac{\partial}{\partial x} + \frac{\partial}{\partial y} + \frac{\partial}{\partial z}$ is the gradient operator, λ is the thermal conductivity of the soil and Q is the heat generated from the heat source. The first term in Eq. (1) represents the change in energy storage with respect to time. The second term in Eq. (1) represents the net energy flow by conduction.

Thermal conductivity is the property of a material to conduct heat and evaluated primarily in terms of Fourier's Law for heat conduction. Soil is a discrete material made up of solid particles. It can be either dry or partly or fully saturated by water. The thermal conductivity of soil depends highly on the type, density and moisture content of the soil material. Consider a partially saturated soil, which contents three phases. Generally speaking, solid have a higher thermal conductivity than liquid, liquid have a higher thermal conductivity than gas. Hence, the thermal conductivity of partially saturated soil increases with the moisture content.

Another important material property related to heat transfer problem is the heat capacity or specific heat. Soil has a specific heat of 800 J/kg/K and water of 4182 J/kg/K. The specific heat of water is much larger than the soil, which means that soil with high moisture content also has high specific heat.

2 Thermal Conductivity Measurements

Several researchers [2–13] have shown that the thermal properties of soil depend on numerous parameters such as mineralogical composition, grain size of soil, and

Table 1 Thermal conductivity of sand

Water content (Weight %)	0.063	0.512	1.727	4.035
Measurement 1	0.2728	0.3116	0.8646	0.9104
Measurement 2	0.2601	0.2766	0.5321	1.0213
Measurement 3	0.2618	0.3499	0.3987	0.9061
Averaged thermal conductivity (W/m/K)	0.2649	0.3127	0.5985	0.9459

Table 2 Thermal conductivity of gravel

Water content (Weight %)	0.080	0.632	1.358	5.034
Measurement 1	0.5272	0.4718	0.2913	0.1815
Measurement 2	0.4215	0.4969	0.4362	0.4075
Measurement 3	0.4552	0.5059	0.2700	0.7475
Averaged thermal conductivity (W/m/K)	0.4680	0.4915	0.3325	0.4455

physical properties like moisture content, porosity, dry density, and degree of saturation. Therefore, these factors have to be taken into account when performing measurements at laboratory and field scale. The two main backfill materials used by Wiener Netze is sand and gravel. This represents a filling material that consists of a graded grain-size distribution. The thermal conductivities of these two soils are measured in the Space Research Institute (Institut für Weltraumforschung, IWF) in Graz. A thermal conductivity sensor thermal needle TP02 is used, which has proven suitability in soils and thermal backfill materials. The thermal needle TP02 is based on the Non-Steady-State Probe (NSSP) measurement method (also known as transient line source, thermal needle, hot needle, heat pulse- and hot wire technique), which has the fundamental advantages that it is fast and absolute while the sample size is not critical. Cylindrical samples with a height of 30 cm and diameter of 25 cm are prepared for the measurement of thermal conductivity. Samples of sand and gravel with different moisture contents are tested. We use dry to medium wet materials, since high moisture content leads to high thermal conductivity which is not interesting for this project. The moisture content in these tests are determined by weighting and drying of the material.

For both sand and gravel, samples with four different moisture contents are prepared. The thermal conductivity of each sample is measured three times and the averaged values are taken. The measurement results for sand is shown in Table 1 and the results of gravel is shown in Table 2.

It can be seen from the tables and figures that the thermal conductivity of sand increases with the water moisture content, which can be expected. The relationship between thermal conductivity and moisture content is proportional.

However, the thermal conductivity of gravel do not have any consistent dependence on the moisture content. Even the measurement results with the same moisture content are very different. The reason could be that the gravel have very divergent particle sizes, which will lead to nonuniform contact between the soil material and the thermal needle. Therefore, we cannot accurately measure the thermal conductivity of gravel. According to the measurement data, we can say that the thermal conductivity of gravel is around 0.45 W/m/K.

3 Model Experiments

The model experiments are designed to reproduce the boundary conditions of the underground cables. The upper layer of the soil is heated to a relatively 50 °C, which is the ground temperature of asphalt road in summer day time.

The length of the structure is much greater than the other two dimensions. The heat transfer associate with length are constrained by nearby cable and soils and are small compared to the cross-sectional heat transfers. The problem is equivalent to plane strain problems, therefore we only need to study the heat transfer in a cross-section of the cable and soil.

A wood box is built to contain the soil sample, which has a hight of 110 cm, width of 90 cm and depth of 20 cm. The soil sample contains a sand layer and a gravel layer, the height of sand layer is 34.5 cm as shown by the dashed line in Fig. 2. A heating pipe is buried 90 cm from the upper surface of the soil in the sand layer. It is used

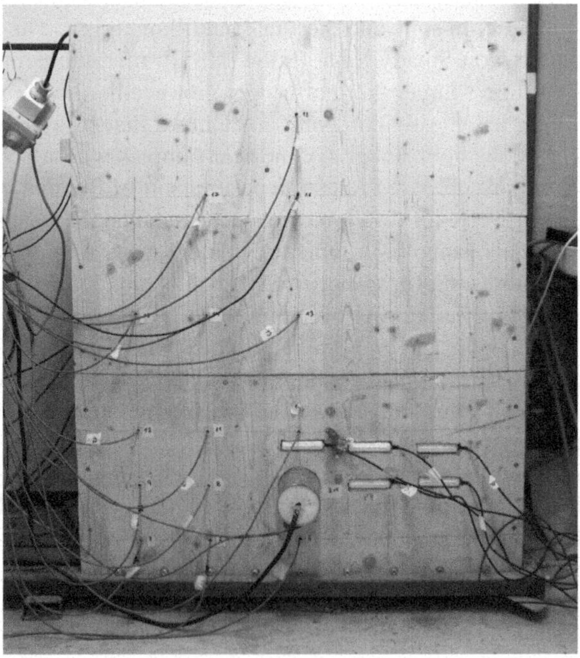

Fig. 1 The photo of the model test

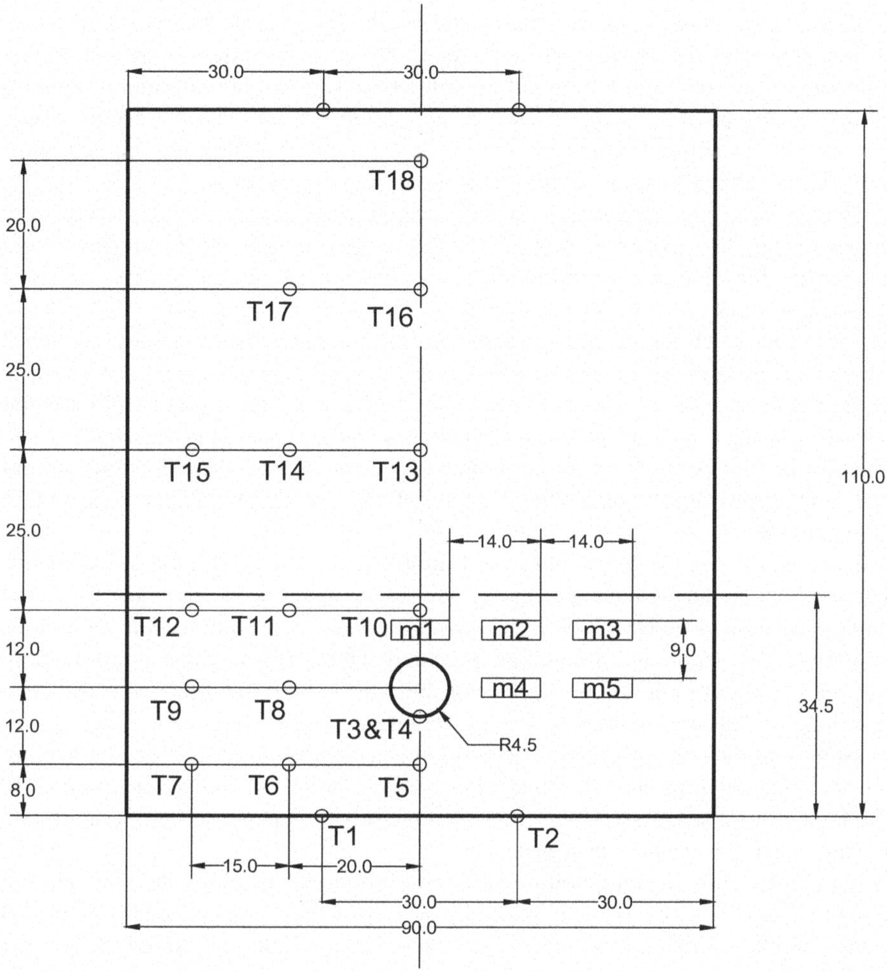

Fig. 2 The size of the model test sample and positions of sensors. T1-T20 are temperature sensors, m1-m5 are moisture sensors, all dimension units are cm. The dashed line shows the interface between sand and gravel

to simulate the cable. Both temperature sensors and soil moisture sensors are placed in the soil sample as shown in Fig. 2, where T1-T20 are temperature sensors and m1 - m5 are moisture tensors. Since the problem is axial symmetric, only half of the model is measured with sensors. The positions of sensors are carefully chosen, so that they can cover the most important regions of the sample while many sensors are placed around the cable. The temperatures on the left and right boundaries of the box should be either constant or the same as in the middle, which is not so interesting. Hence, no temperature sensors are placed on the left and right boundaries. The photo of the model box is shown in Fig. 1.

The temperatures of soil depend on the depth. The upper surface is hot due to the sunshine and the lower surface is relatively cool. In order to reproduce this temperature difference in the model experiments, we use a heating mat to heat up the the upper part of the soil and a water circulation system to cool down the lower part of the soil sample. In our experiments the lower layer of the soil is cooled down to 16 °C, which is the initial condition of the model experiment.

The GS3 soil moisture sensors from Decagon are used to measure the moisture change during the model experiment. The GS3 sensor uses an electromagnetic field to measure the dielectric permittivity of the surrounding medium. Five GS3 soil moisture sensors are used in the model experiments. The position of sensors are shown in Fig. 2. All the moisture sensors are calibrated so that the measured values can be converted to the weight percentage of the sand material we used. Generally speaking, an increase of temperature in soils will lead to decrease of soil moisture contents (drying). This is consistent with what we have observed in the experiments: when the box is opened after the experiments, we can see that the color soils around the heating pipe is lighter than other the color of soils, i.e. there is a dried zone around the heating pipe.

The heating mat on the top of the soil sample and the heating pipe in the bottom will first heat up during the experiments. We set the upper heating mat to 50 °C and the heating pipe to be 60 °C or 75 °C. For each case, three sand samples with low, medium and high moisture contents are prepared, which correspond to approximately 2, 4 and 6% water content by weight. For the gravel layer, materials with the same moisture content (approximately 4% by weight) are used. Altogether, six groups of measurement data are obtained. The soil is heated for about 24 h, then the heating mat and heating pipe are switched off to let the soil cool down for about 24 h. The temperature sensors and the moistures records the change of temperature and moisture during the whole process.

In order to study the temperature variations around the heating pipe, the measure temperatures of sensors 3–15 are plotted. For heating pipe temperature 60 °C of soil sample with low, medium and high water contents, the results are plotted in Figs. 3, 4 and 5. For heating pipe temperature 75 °C of soil sample with low, medium and high water contents, the results are plotted in Figs. 6, 7 and 8. In all the plots, the dotted line defines the time boundary between heating and cooling. From these figures, it can be seen that the heating pipe only lead to the temperature changes in a localized zone. Sensors about 30 cm away from the heating pipe do not show large temperature change. Since the heating period in the experiments is about 24 h, the affected zone is even smaller for underground cables.

In addition, several tests with complete dry soils are carried out. The temperatures of the heating pipe are set to be 50, 60, 70 and 80°C for these tests. The results are shown in Fig. 9. Only the measurement of sensor T5, T3/T4 and T10 are shown, since other sensors do not show large temperature change. As can be seen from the plots, the heat dissipation in dry soil is much slower that in moist soils. Comparing the measurements of sensors 5 and 10, it can be seen that high heating pipe temperature

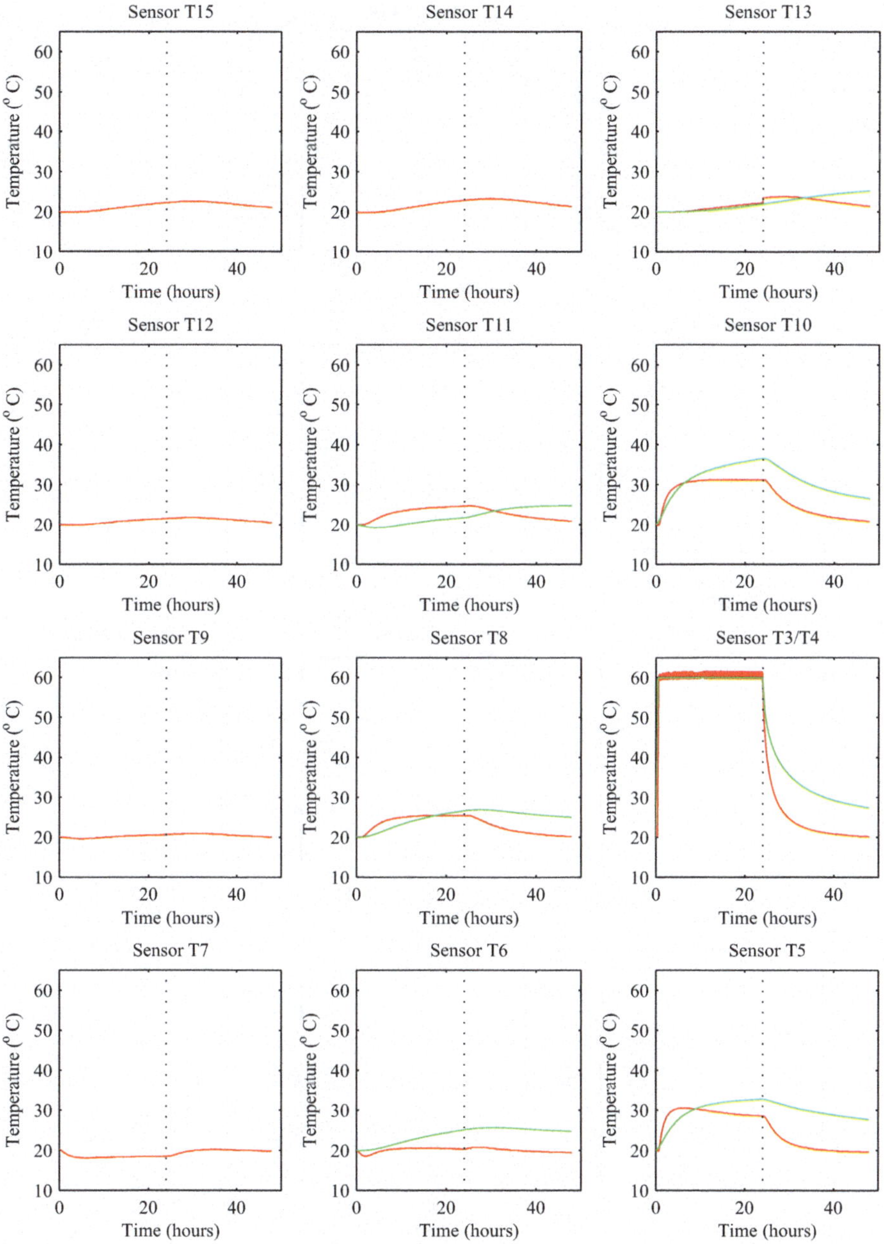

Fig. 3 Temperature sensor measurements (red) and FEM results (green) for heating pipe temperature 60 °C and soil moisture content 2%

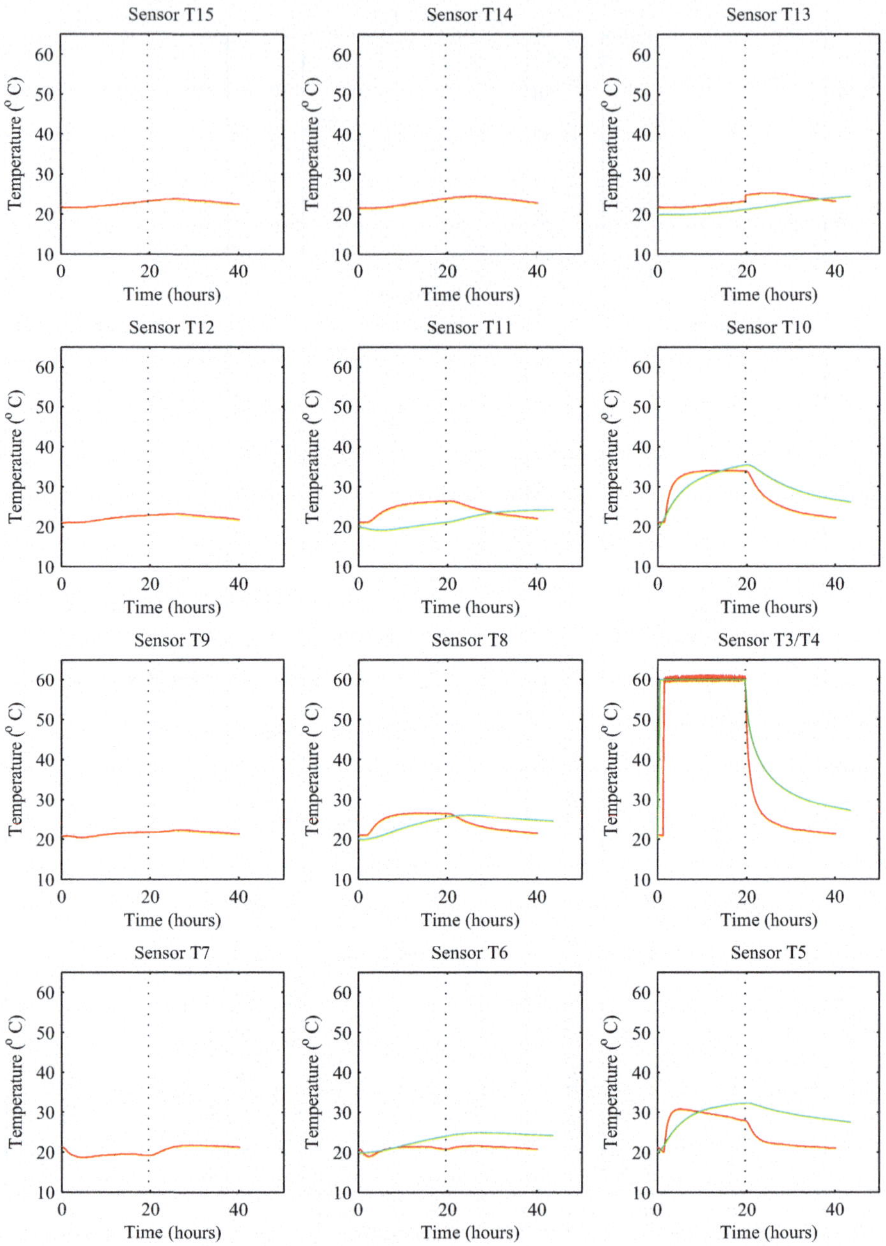

Fig. 4 Temperature sensor measurements (red) and FEM results (green) for heating pipe temperature 60 °C and soil moisture content 4%

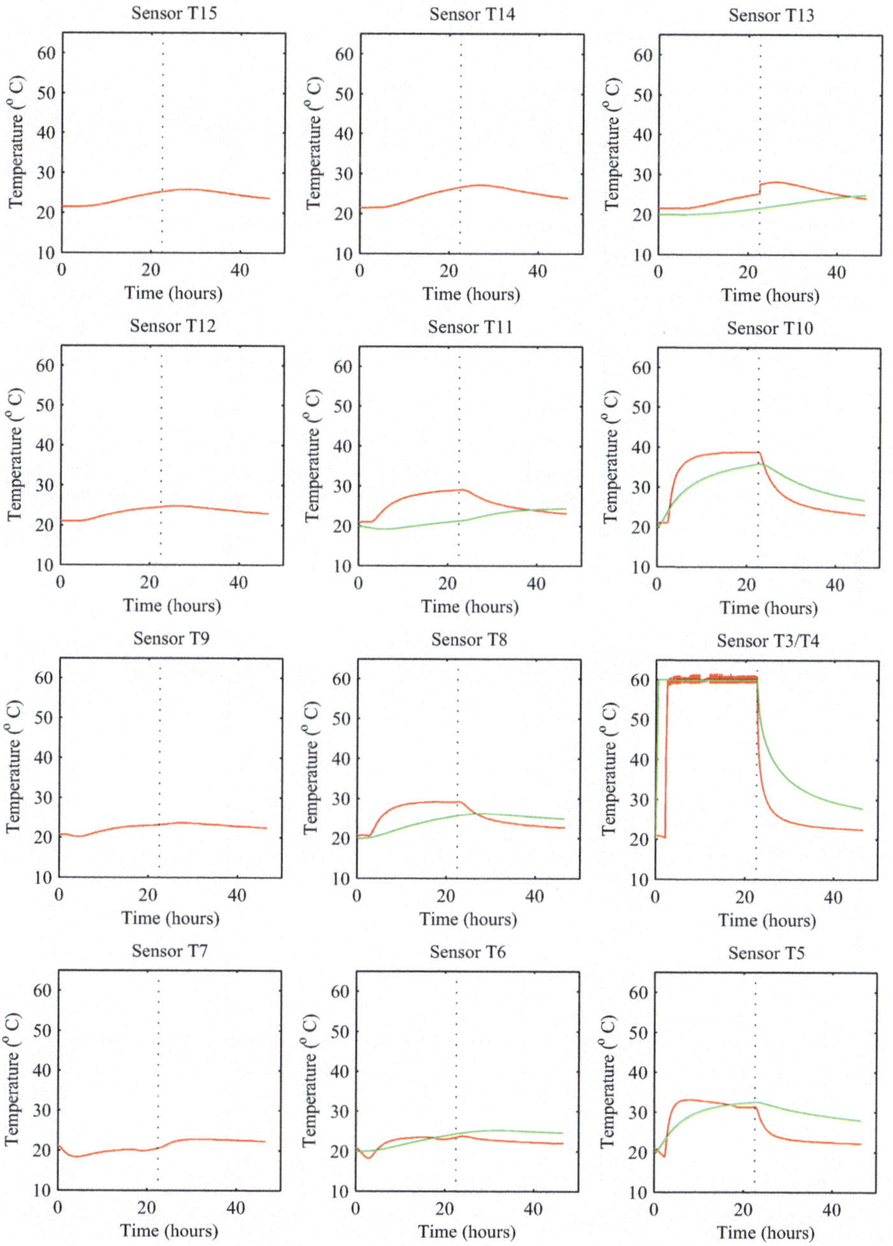

Fig. 5 Temperature sensor measurements (red) and FEM results (green) for heating pipe temperature 60 °C and soil moisture content 6%

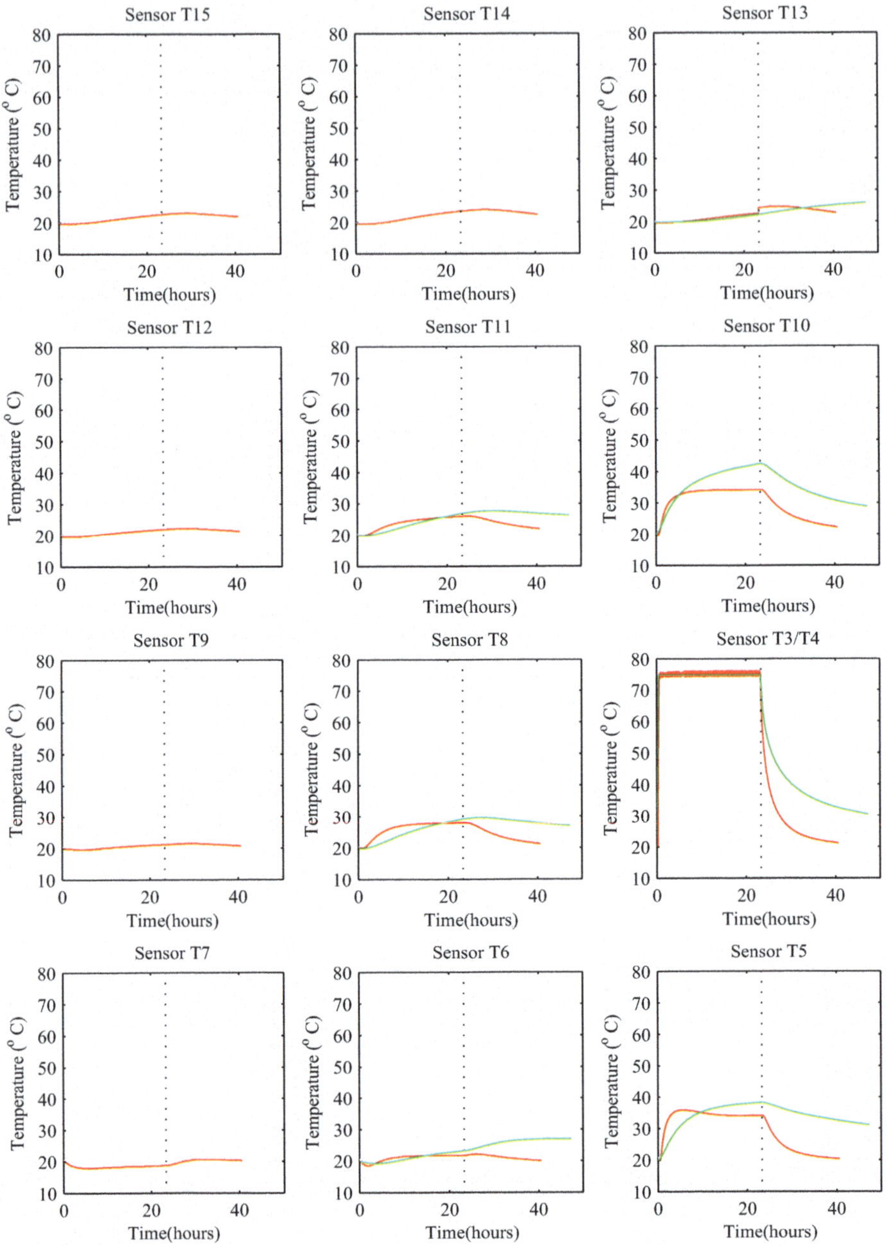

Fig. 6 Temperature sensor measurements (red) and FEM results (green) for heating pipe temperature 75 °C and soil moisture content 2%

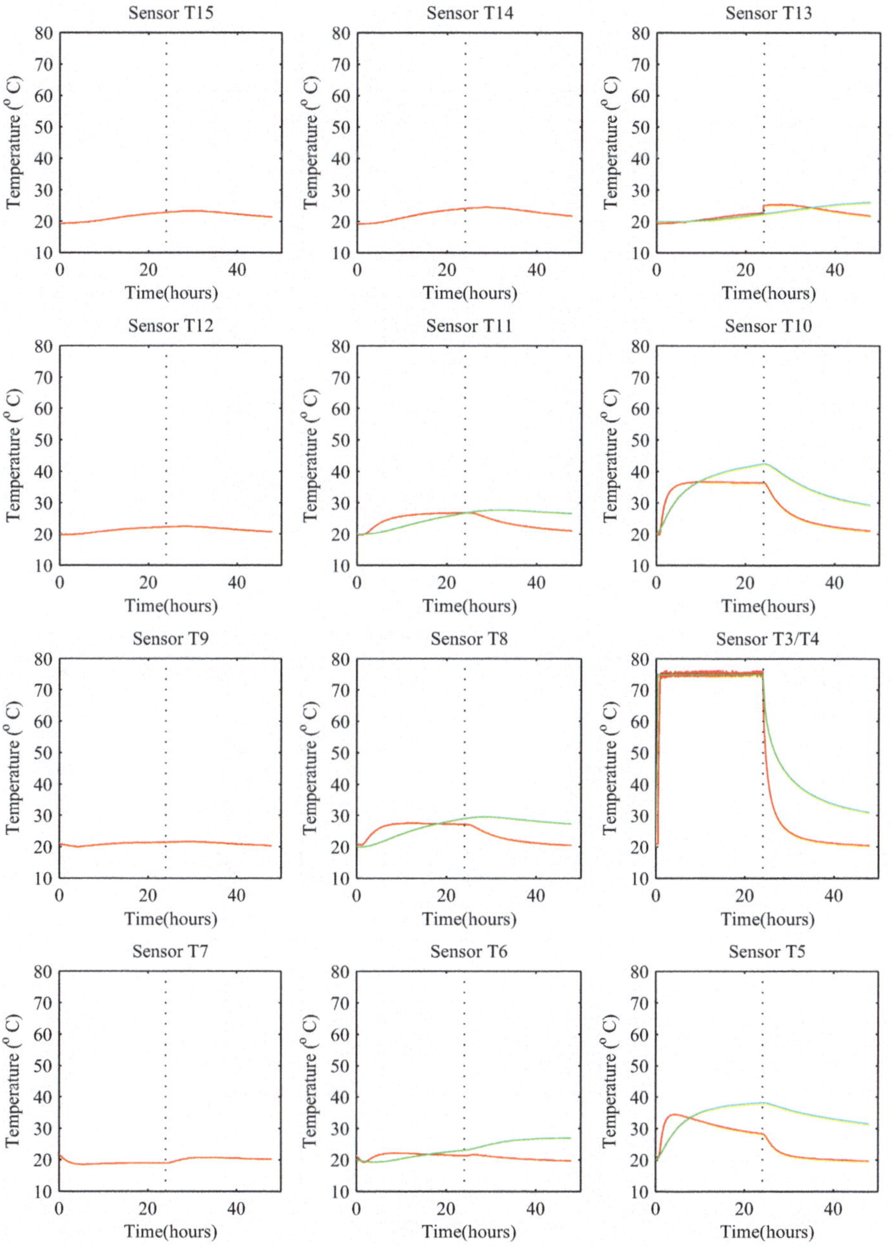

Fig. 7 Temperature sensor measurements (red) and FEM results (green) for heating pipe temperature 75 °C and soil moisture content 4%

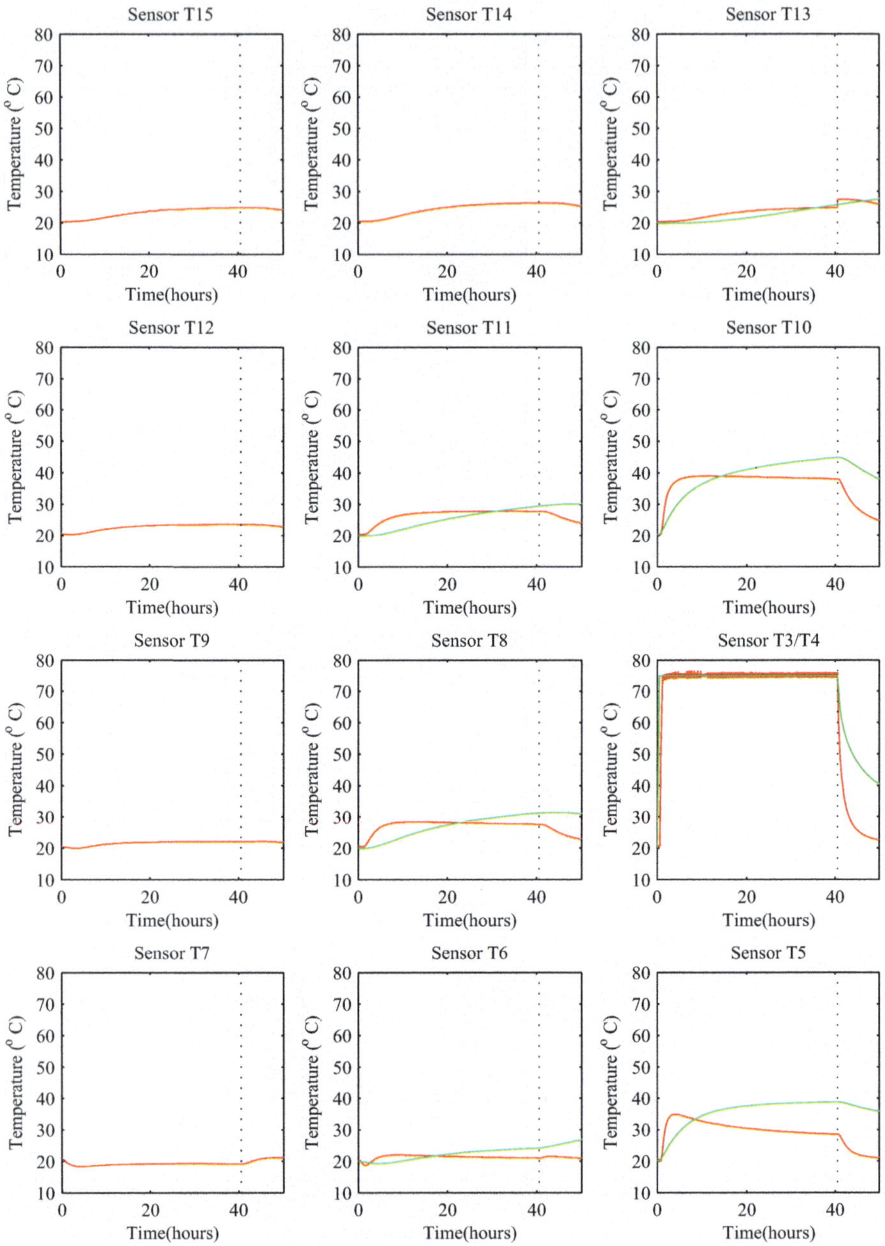

Fig. 8 Temperature sensor measurements (red) and FEM results (green) for heating pipe temperature 75 °C and soil moisture content 6% labelfig

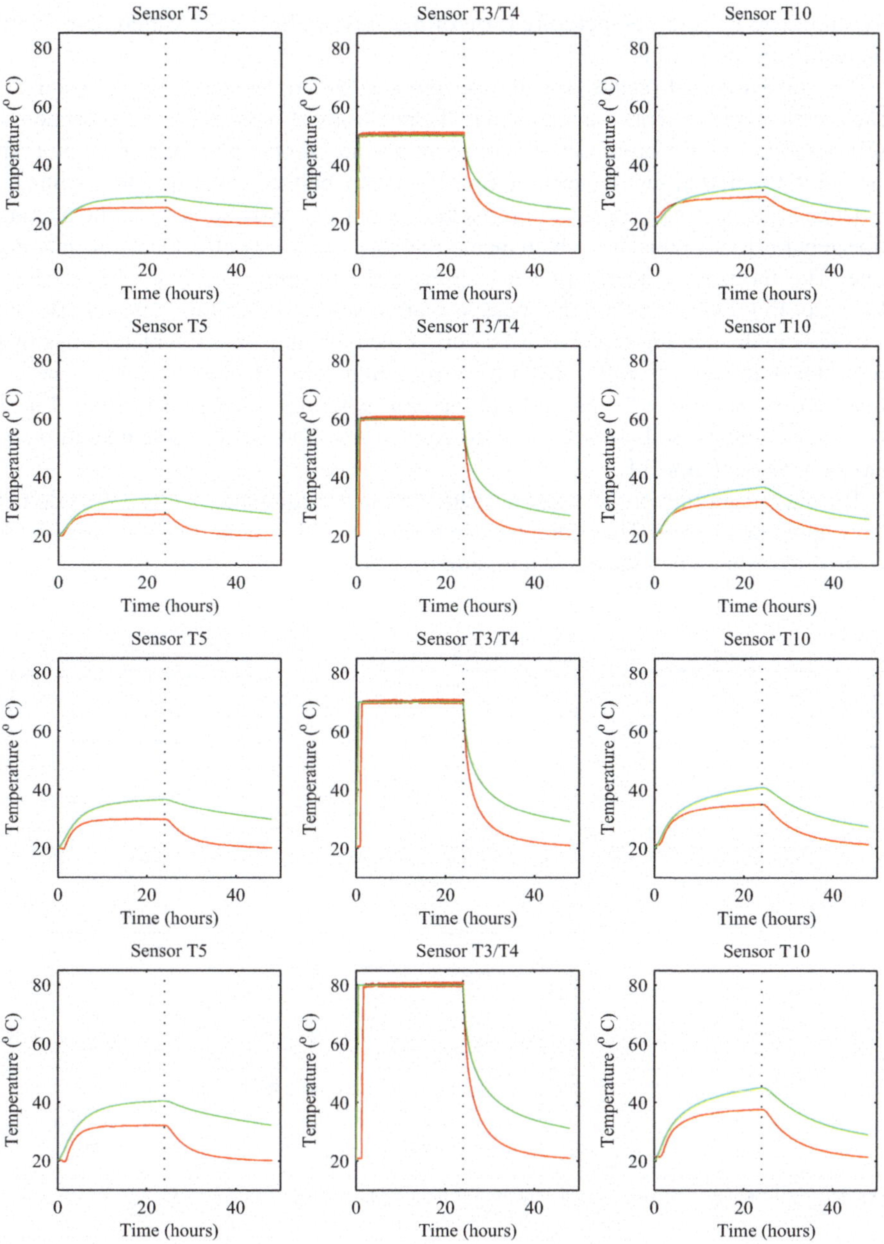

Fig. 9 Temperature sensor measurements (red) and FEM results (green) for dry soil with different heating pipe temperatures

does not lead to great temperature raise of the surrounding soils, which agree with the results in moist soils.

The measurement results of soil moisture sensors are shown in Figs. 10 and 11. The positions of the plots are identical to the positions of sensors in the experiments. Soil samples with different initial water contents are plotted with different colors in each plot, the dotted line defines the time boundary between heating and cooling.

During the heating process, the moisture contents of soils away from the heating pipe generally stay constant, which means that they are barely affected by the heating pipe. The moisture contents of soils close to the heating pipe first increase then decrease. In order to explain the measurement results, we need to consider that the soils next to the heating pipe are dried during the heating process, which means that the water in these soils will transfer to surrounding soils. Hence, the water contents measured by sensors m1, m4 and m2 increase when the pipe start heating. Later, the water contents in these soils will decrease due to the raising of temperature, as shown in Figs. 10 and 11.

During the cooling process, the measured moisture contents for all sensors do not show apparent change. Therefore, we can assume that the water do not transfer back to the dried zone when the cable temperature is low.

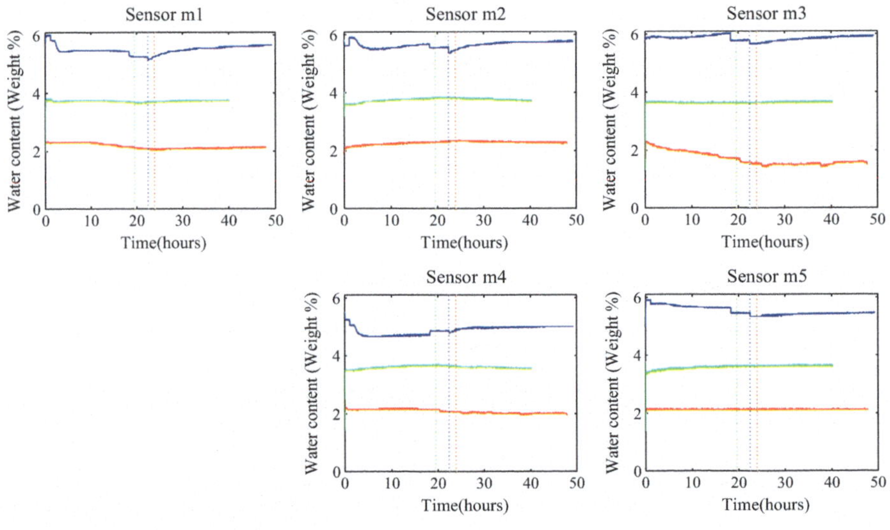

Fig. 10 Soil moisture sensor measurements for heating pipe temperature 60 °C.

4 Numerical Simulations

In this section, we analyze heat transfer problem of underground cable with numerical methods. The finite element analysis software ABAQUS is used. As mentioned in the above sections, the problem is plain strain. Hence, we use 2-D plane strain elements to carry out the simulations. The numerical simulation model has the same dimension as the model experiment, which is 110 cm high and 90 cm wide. The FEM mesh contents 11543 4-node linear heat transfer quadrilateral elements with a size around 1 cm × 1 cm. The initial temperature is 20 °C throughout the entire soil mass. During heating, the temperature at the upper boundary is set to be 50 °C and the temperature at the lower boundary is set to be 16 °C. The temperature of the cable is either 60 °C or 75 °C.

The model is divided into two layers, as in the model test, the upper layer is gravel with a height of 75.5 cm, the lower layer is sand with a height of 34.5 cm. The cable is buried in the sand layer. The gravel layer has a density of 1600 kg/m^3, specific heat of 950 J/kg/K and thermal conductivity of 0.45 W/m/K. For the sand layer, since different water contents are considered, material parameters in Table 3 can be calculated based on the water content and the measurement results of Sect. 2.

In order to have a direct comparison between the FEM and experimental results, the temperatures obtained from FEM simulations are also plotted in Figs. 3–9. Nodes in the positions of sensors T3-T6, T8, T10, T11, T13 are taken from the FEM model, the temperatures at these nodes over time are plotted. It can be seen that the experimental results can be well modeled with FEM simulations. However, in the heating

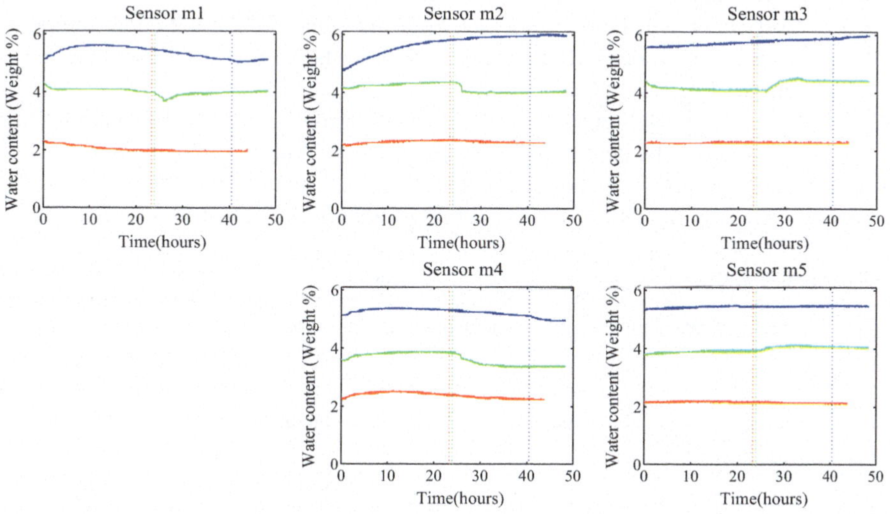

Fig. 11 Soil moisture sensor measurements for heating pipe temperature 75 °C

period, the temperatures predicted by FEM simulations are generally a bit higher than the temperatures measured by sensors. In the cooling period, the FEM simulations show slower cooling rate than in the experiments. The reason could be due to the loss of heat in the out-of-plane direction during the experiments. A plane strain condition is assumed for the experiments, which means that there is no heat transfer in the out-of-plane direction. However, it is inevitable to have heat loss in the out-of-plane directions if the model temperature is higher than the room temperature.

5 Conclusion

The experiments and numerical simulations presented in this paper show the following conclusions:

- It is difficult to determine the thermal conductivity of gravel. For sand, the thermal conductivity has a linear dependency on the water content.
- The heating on the upper surface of the soil has an influence depth of about 30–40 cm. If the cable is buried deep enough, the influence of the upper surface heating (e.g. solar radiation) can be neglected.
- Since the temperature of the soil decrease with the depth, the upper part of the cable will have higher temperature than the lower part. The highest temperature will be on the upper point of the cable.
- The heat dissipation depends on the moisture content of surrounding soils. High cable temperature does not affect the heat dissipation rate very much.
- During the heating of the underground cable, soils around the cable will be dried. Soil water will be driven away from the cable. When the cable temperature decrease, soil water will not transfer back to the soils around the cable, i.e. a dried zone will be formed around the cable.
- With the material parameters measured from the experiments, the FEM modelings are able to predict the heat transfer problems of the underground cable.
- The model test trends to underestimate the temperatures of the soil and overestimate the cooling rate, since there is a loss of heat in the out-of-plane direction.

Acknowledgements This research was supported by the city of Vienna and the company Wiener Netze. We thank M. Schort, Mr. Bolzer, Mr. Gottsbacher from Wiener Netze for assistance with data and technology. We would also like to show our gratitude to the Prof. Korjenic from institute of

Table 3 Material parameters for sand with different water contents

Water content (weight %)	0	2	4	6
Density (kg/m^3)	1520	1551	1583	1617
Specific heat (J/kg/K)	800	868	935	1003
Thermal conductivity (W/m/K)	0.2525	0.6035	0.9545	1.3055

building construction and technology Vienna university of technology for sharing her experiences in measuring soil temperature.

References

1. Reichl, J., Schmidthaler, M., Schneider, F.: The value of electricity supply security: The costs of power outages to households, firms and the public sector. Energy Econ. **36**, 256–261 (2013)
2. de Vries, D.A.: Thermal properties of soil, Physics of Plant Environment, pp. 210–235 (1963)
3. Kersten, M.S.: Thermal properties of soils, Bulletin 28. University of Minnesota, Minneapolis, MN, USA, Engineering Experiment Station (1949)
4. Penner, E., Johenston, G.H., Goodrich, L.E.: Thermal conductivity laboratory studies of some Mackezie highway soils. Can. Geotech. J. **12**, 271–288 (1975)
5. Salomone, L.A., Kovacs, W.D., Kusuda, T.: Thermal performance of fine-grained soils. J. Geotech. Eng. **110**(3), 359–374 (1984)
6. Salomone, L.A., Marlowe, J.I.: Soil rock classification according to thermal conductivity, EPRI CU-6482. Electric Power Research Institute, Palo Alto, CA, USA (1989)
7. Brandon, T.L., Mitchell, J.K.: Factors influencing thermal resistivity of sands. J. Geotech. Eng. **115**(12), 1683–1698 (1989)
8. Mitchell, J.K.: Conduction phenomena: from theory to geotechnical practice. Geotechnique **41**(3), 299–340 (1991)
9. Tarnawski, V.R., Leong, W.H.: Thermal conductivity of soils at very low moisture content and moderate temperatures. Transp. Porous Media **41**(2), 137–147 (2000)
10. Campbell, G.S., Norman, J.M.: An Introduction to Environmental Biophysics, 2nd edn. Springer, New York (1998)
11. Rao, M.V.B.B., Singh, D.N.: A generalized relationship to estimate thermal resistivity of soils. Can. Geotech. J. **36**(4), 767–773 (1999)
12. Singh, D.N., Devid, K.: Generalized relationships for estimating soil thermal resistivity. Exp. Therm. Fluid Sci. **22**(3–4), 133–143 (2000)
13. Naidu, D., Singh, D.N.: A generalized procedure for determining thermal resistivity of soils. Int. J. Therm. Sci. **43**(1), 43–51 (2004)

Analytical Solutions for Steady Granular Flows in Simple Configurations

Chong Peng, Xiaogang Guo and Wei Wu

Abstract Most granular materials encountered in nature and industry lie either in the quasi-static regime or the intermediate dense flow regime. A unified constitutive model for granular material in the quasi-static and dense flow regimes is introduced. The main granular particles interactions controlling the mechanical behaviors, frictional contact and collision, are taken into account using the Mohr-Coulomb model and a Bagnold-type relation, respectively. Granular flows in three simple configurations, i.e., plain shear, vertical chute flow and flow on an inclined plane, are studied. Analytical solutions based on the presented unified model are obtained. Comparison between results from the presented model and the $\mu(I)$ model indicates that the explicit partition of frictional and collisional stress components provides insights in dense granular flows.

1 Introduction

Granular flows widely present in industry and nature, e.g., raw material processing and geophysical flows. With different material properties, external loads and boundary conditions, granular materials can behave like solid, fluid or gas. In the quasi-static regime, the solid-like granular materials can be sufficiently described using plasticity theories in soil mechanics, whereas constitutive equations based on

C. Peng (✉) · X. Guo · W. Wu
Institute of Geotechnical Engineering, University of Natural Resources and Life Sciences, Vienna, Feistmantelstrasse 4, 1180 Vienna, Austria
e-mail: pengchong@boku.ac.at

X. Guo
e-mail: xiaogang.guo@boku.ac.at

W. Wu
e-mail: wei.wu@boku.ac.at

W. Wu (ed.), *Recent Advances in Geotechnical Research*,
Springer Series in Geomechanics and Geoengineering,
https://doi.org/10.1007/978-3-319-89671-7_13

kinetic theory have been proposed for the gas regime [1]. However, most granular flows lie neither in the slow shearing quasi-static regime, nor the fully collisional gas regime. In the intermediate flow regime, the granular volume fraction is close to its maximum value; thus, the flows can be considered incompressible. This flow regime is usually referred to as dense flow regime in literature.

In the dense flow regime, the granular material flows like a liquid. Two distinct features are observed for dense granular flows. The first is that there exists a threshold yield stress below which the granular media stops flowing; the second is that the granular dense flows are shear-rate-dependent. These two features arise from the two controlling particle interactions in dense granular flows: frictional contact and collision. The continuously evolving network of frictional contacts result in the frictional behaviors in the dense granular flows, whereas the viscous rate-dependent behavior has its source in particle collisions. Which interaction is more dominant depends on several factors, e.g., material property, external load and boundary condition. A widely accepted constitutive model for the dense granular flow regime is the $\mu(I)$ model [2]. It employs an overall phenomenological frictional coefficient related to a dimensionless 'inertial number'. Frictional contact and collision are not explicitly considered in this model. Moreover, the mechanical behaviors inside the yield surface is undefined; thus, the $\mu(I)$ model cannot be applied to granular materials in the quasi-static regime.

Recently, a unified constitutive framework for granular media in the quasi-static and dense flow regimes is proposed [3, 4]. The two controlling interactions, frictional contact and collision, are taken into account using a rate-independent hypoplastic model and the Bagnold-type relation, respectively. It was demonstrated that the framework provides good modeling for granular materials under slow plastic deforming and fast collisional shearing. In this work, granular flows in simple configurations are considered. Analytical solutions based on the presented unified model are derived. The results are compared with the analytical solutions obtained using the $\mu(I)$ model. The advantages and unsolved matters of the unified model are discussed.

2 Unified Model for Granular Media

In the quasi-static regime, the inter-particle frictional forces result in rate-independent Coulomb-type plastic behaviors. In the dense flow regime, the inter-particle frictional contacts still contribute to the development of stress components. However, a notable part of stress comes from the moment transfer in inter-particle collision. A unified model should consider these two interactions. Consequently, the following stress form is assumed in the unified model

$$\sigma = \sigma^{frictional} + \sigma^{collisional} \tag{1}$$

where σ is the Cauchy stress tensor, $\sigma^{frictional}$ and $\sigma^{collisional}$, later referred as σ^f and σ^c, are two stress parts originating from frictional contact and collision.

Constitutive models for cohesionless geomaterials can be chosen to describe the frictional behaviors of granular materials.

Common properties of geomaterials such as nonlinear elastic response, plastic deformation and critical state, can be found in granular materials in the quasi-static regime. The rate-independent constitutive component for the unified model can be chosen according to the requirement that how comprehensively it is required in modeling the behaviors inside the yield surface. For simplicity, the Mohr-Coulomb model is applied for the quasi-static regime in this work, because we only focus on steady granular flows. When medium yields at a point, there exists a plane on which the stress satisfy

$$| \tau^f | = \sigma_n^f \tan \phi \tag{2}$$

where σ_n^f and τ^f are the normal and shear stress on the shear plane.

In fast shearing where the particle collision dominates the bulk behavior and dissipation of kinetic energy, experiments and theoretical analysis suggest a quadratic dependence of shear stress on shear strain rate in the simple shear configuration [6]

$$\tau_{xz}^c = k_v \left(\frac{\partial v_x}{\partial z} \right)^2 \tag{3}$$

where k_v is a coefficient related to particle size, density and volume fraction; x and z are Cartesian coordinates. Moreover, a collision induced pressure p^c, termed dispersive pressure, is found proportional to the shear stress in rapid shearing [6]

$$p^c = \tau_{xz}^c / \tan \alpha \tag{4}$$

where α denotes the ratio between the shear stress and dispersive pressure in the rapid shearing. It is found that α is a constant for granular media.

In general granular flows where particle interactions possibly include friction and collision, both stress parts take effect simultaneously. The proportions of frictional and collisional stresses depend on the kinematics, loading and boundary conditions of flows. In very slow quasi-static deforming, the collisional part is negligibly small. The mechanical response of the granular media is plastic and can be modeled using the Mohr-Coulomb model. From very slow motion to rapid flow, the dynamic part becomes more and more significant. This results in an increase in the dispersive pressure p^c and a tendency of dilation, which leads to a relaxation in the normal stress of frictional contact. Therefore, the rate-independent stress part tends to decrease with the increase of shear rate. In very fast shearing, the particles are agitated and interact mainly by binary collisions. The network of frictional contacts no longer exists in this stage, and the granular media are in the gaseous regime, not the dense flow regime.

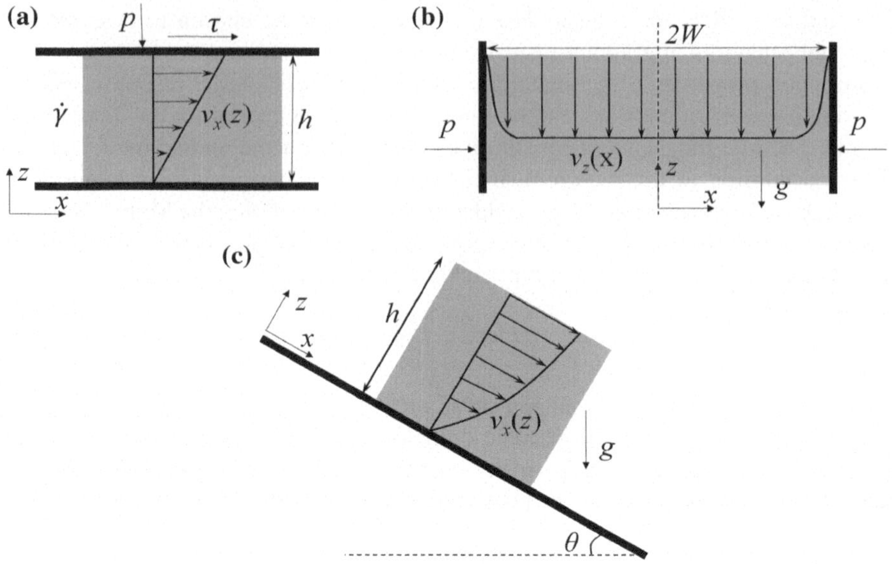

Fig. 1 The geometry and boundary conditions in the three granular flow problems

3 Analytical Solutions for Steady Granular Flows

Three granular flows under simple configurations are studied. Analytical solutions are obtained and compared to the solutions from the $\mu(I)$ model. The geometries and boundary conditions of the three granular flows are shown in Fig. 1.

3.1 Plane Shear

Consider shearing of a granular material made of particles with diameter d and density ρ_p, as shown in Fig. 1a. The confining pressure p is imposed on the top plate. The granular material is sheared at a prescribed rate $\dot{\gamma} = v/h$, where v is the velocity of the top plate and h is the depth of the granular body. Gravity is not considered. To maintain a steady plane shear, the shear force on the top plate depends on the imposed confining pressure and shear rate.

In the $\mu(I)$ model, the inertial number can be calculated through

$$I = \frac{\dot{\gamma} d}{\sqrt{p/\rho_p}} \tag{5}$$

Consequently, we can obtain the shear force τ by [2]

$$\tau = \mu(I)p, \text{ with } \mu(I) = \mu_1 + \frac{\mu_2 - \mu_1}{I_0/I + 1} \quad (6)$$

where μ_1, μ_2 and I_0 are three parameters of the $\mu(I)$ model.

With the presented unified model, considering the steady shearing, we have the following two equations

$$\tau = \tau^f + \tau^c = \sigma_{zz}^f \sin\phi + k_v\dot{\gamma}^2 \quad (7)$$

$$p = \sigma_{zz}^f + p^c = \sigma_{zz}^f + k_v\dot{\gamma}^2/\tan\alpha \quad (8)$$

The frictional shear force is computed as $\sigma_{zz}^f \sin\phi$, not $\sigma_{zz}^f \tan\phi$, because simulations of plane shear indicate that the difference between the vertical and the horizontal normal stress is very low, i.e., $\sigma_{xx}^f \approx \sigma_{zz}^f$ [7, 8]. The frictional shear stress is then equal to $\sigma_{zz}^f \sin\phi$. Based on Eqs. (7) and (8), the total shear force is obtained as

$$\tau = p\sin\phi + k_v\dot{\gamma}^2(1 - \frac{\sin\phi}{\tan\alpha}) \quad (9)$$

Comparing Eqs. (6) and (9), it can be observed that the results of shear stress obtained using the $\mu(I)$ model and the unified model have a similar structure. Both results consist of two parts, one part is constant with given normal stress p, and the other one depends on the shear rate. In the dense flow regime, the shear stress obtained using the $\mu(I)$ model ranges from $\mu_1 p$ to $\mu_2 p$. On the other hand, in the dense flow regime, the dispersive pressure $p^c = k_v\dot{\gamma}^2/\tan\alpha$ is from 0 to p; therefore, the range of shear stress obtained using the unified model is from $p\sin\phi$ to $p\tan\alpha$. It is observed that the constants $\sin\phi$ and $\tan\alpha$ in the unified model are equivalent to the constants μ_1 and μ_2 in the $\mu(I)$ model.

Both of the models enable precise classifications of the three flow regimes: quasi-static, dense and gaseous. Small values of $\dot{\gamma}$ ($\dot{\gamma} \to 0$) correspond to the quasi-static regime in the sense that deformation is very slow. In the $\mu(I)$ model, the threshold of I corresponds to the rapid and dilute flow regime is $I \approx 1$, which is obtained from experiments [8]. Contrarily, in the unified model, the gaseous regime can be theoretically determined: it corresponds to the state when the network of frictional contacts disappears, that is, when $p^f = 0$. Based on the aforementioned analysis, the shear rate corresponds to the dense flow regime in the plane shear is

$$\begin{aligned} &\text{For the } \mu(I) \text{ model}: \ 0 < \dot{\gamma} \lesssim \sqrt{p}/\sqrt{\rho_p d^2} \\ &\text{For the unified model}: \ 0 < \dot{\gamma} < \sqrt{p}/\sqrt{k_v} \end{aligned} \quad (10)$$

Again, the ranges of shear rate corresponding to the dense flow regime obtained using the $\mu(I)$ model and the unified model have similar form. Note that the constant k_v is related to particle diameter, density and volume fraction. It can be observed that in both of the models, the upper limits of the shear rate depend on the confining pressure p and physical properties of the granular material.

Although these two models bears some similarities, there are fundamental differences. The $\mu(I)$ model employs an inertial number-dependent phenomenological frictional coefficient, which changes with shear rate and confining pressure. However, the model does not give any clue why the phenomenological frictional coefficient $\mu(I)$ changes with different shear rate and confining pressure. On the other hand, the unified model explicitly considers the origin of different stress components. In the unified model, the change of phenomenological frictional coefficient can find its explanation as follows. The ratio between shear stress and normal stress in the network of frictional contacts is related to the internal friction angle ϕ of the material, whereas the shear/normal ratio of the collision-induced stresses is related to another angle α. Experiments show that α is generally larger than ϕ. Therefore, when the flow state moves from the quasi-static regime to the gaseous regime, the collision-induced stresses becomes more and more significant. In this process, the phenomenological frictional coefficient increases from $\sin\phi$ to $\tan\alpha$. The evolution of the phenomenological frictional coefficient in the two models is written as

$$\begin{aligned} &\text{In the } \mu(I) \text{ model}: \ \mu = \mu_1 + \frac{\dot{\gamma}d(\mu_1 - \mu_2)}{I_0\sqrt{p/\rho_p} + \dot{\gamma}d} \\ &\text{In the unified model}: \ \mu = \sin\phi + \frac{\dot{\gamma}^2 k_v(\tan\alpha - \sin\phi)}{p\tan\alpha} \end{aligned} \tag{11}$$

Comparing Eq. (11) with Eq. (6), it is observed that the evolution of the phenomenological frictional coefficient in the two models has similar structure, but different dependence on shear rate and confining pressure.

3.2 Vertical Chute Flow

Consider a granular material flowing between two vertical plates with a distance of $2W$, as shown in Fig. 1b. The confining pressure on the two vertical plates is p, which implies that $\sigma_{xx} = p$ within the flow in the steady state.

According to the unified model, the possible range of shear stress τ_{xz} in the chute flow is from $p\sin\phi$ to $p\tan\alpha$. Because of symmetry, the shear stress at the center of the flow is zero. Based on force balance along the x axis, the shear stress can be written as

$$\tau_{xz} = \rho g x \tag{12}$$

where ρ is the bulk density of the granular material. The steady flow condition requires that the shear stress increases linearly from the flow center to the vertical plates. Consequently, there exists an unsheared zone in which $\tau_{xz} < p\sin\phi$. Namely, when $x < x_p$ ($x_p = p\sin\phi/\rho g$), the granular material is unyielded and moves as a plug layer. If $x_p > W$, the whole granular body is unyielded and no granular flow is possible. On the other hand, we should have that $\tau_{xz} < p\tan\alpha$, i.e., the maximum shear stress required for the force balance does not exceed the maximum shear stress that can be provided. Therefore, according to the unified model, steady granular flow is possible in the vertical chute only if

$$\rho g W/\tan\alpha < p < \rho g W/\sin\phi \tag{13}$$

Now we proceed to derive the velocity profile of the steady flow. The force balances in the vertical and horizontal directions are written as

$$\begin{aligned} p &= \sigma^f_{xx} + \sigma^c_{xx} = \sigma^f_{xx} + \frac{k_v}{\tan\alpha}\left(\frac{\mathrm{d}v_z}{\mathrm{d}x}\right)^2 \\ \tau_{xz} &= \tau^f_{xz} + \tau^c_{xz} = \sigma^f_{xx}\sin\phi + k_v\left(\frac{\mathrm{d}v_z}{\mathrm{d}x}\right)^2 \end{aligned} \tag{14}$$

Substituting Eq. (14) into Eq. (12), we have

$$\left(\frac{\mathrm{d}v_z}{\mathrm{d}x}\right)^2 = \frac{(\rho g x - p\sin\phi)\tan\alpha}{k_v(\tan\alpha - \sin\phi)} \tag{15}$$

Solving the above equation using the boundary condition $v_z = 0$ at $x = W$, we can obtain the velocity profile

$$v_z(x) = v_z^p + \frac{2\sqrt{\tan\alpha}(\rho g x - p\sin\phi)^{3/2}}{3\sqrt{k_v}(\tan\alpha - \sin\phi)\rho g} \quad \text{if } x_p \le x \le W \tag{16}$$

where v_z^p is the velocity of the unsheared plug layer

$$v_z^p = -\frac{2\sqrt{\tan\alpha}(\rho g W - p\sin\phi)^{3/2}}{3\sqrt{k_v}(\tan\alpha - \sin\phi)\rho g} \quad \text{if } x < x_p \tag{17}$$

Analytical solutions can also be obtained using the $\mu(I)$ model [5]. Similarly, an unsheared plug layer exists, and there is a confining pressure range, only with confining pressure in this range a steady granular flow is possible. If the parameters in the $\mu(I)$ model are taken as $\mu_1 = \sin\phi$ and $\mu_2 = \tan\alpha$, the width of the plug layer and the range of the confining pressure are identical to those obtained using the unified model. However, the velocity profiles obtained using these two models are not identical.

To quantitatively compare the solutions obtained using these two models, the velocity profiles are calculated. The following material properties are used: $\rho = 1470\,\mathrm{kg/m^3}$, $\rho_p = 2450\,\mathrm{kg/m^3}$ and $d = 0.002\,\mathrm{m}$. The parameters for the $\mu(I)$ model are: $\mu_1 = 0.38$, $\mu_2 = 0.64$ and $I_0 = 0.279$. For the unified model, we take $\sin\phi = \mu_1$, $\tan\alpha = \mu_2$ and $k_v = 0.2$. The width of the chute is 0.2 m, the confining pressure p is 3000 Pa. Owing to the symmetric geometry, only the right half of the granular material is considered. The calculated results are shown in Fig. 2. Using the given

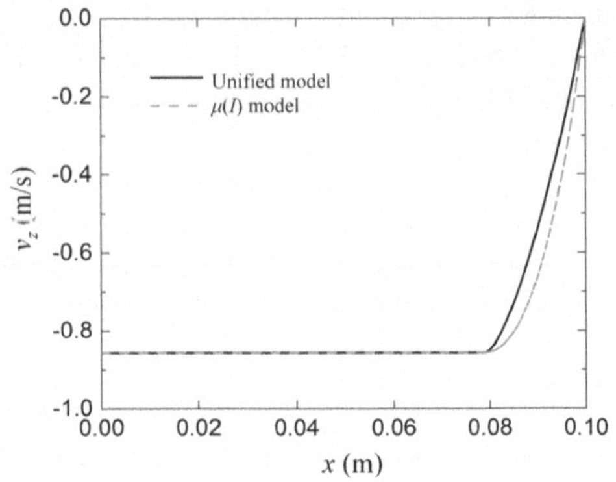

Fig. 2 The velocity profiles obtained using the unified model and the $\mu(I)$ model in the vertical chute flow

parameters, the two models predict identical unsheared velocity. However, the velocity evolution in the shear zone is different. Compared with the results obtained using the $\mu(I)$ model, the velocity obtained using the unified model decreases faster near the plug layer, and slower near the vertical plate.

3.3 Flow On an Inclined Plane

The problem of a granular mass on an incline is considered. As shown in Fig. 1c, the granular mass has a free surface and is subjected to gravity. The inclination is θ. The height of the granular layer is h.

For steady granular flow, the stress components at arbitrary height should be hydrostatic

$$\sigma_{zz} = \rho g(h - z)\cos\theta, \quad \sigma_{xz} = \rho g(h - z)\sin\theta \tag{18}$$

For an infinite inclined plane, we assume that $\sigma_{zz}^{f} = \sigma_{xx}^{f}$ in the flowing state; thus, the shear/normal stress ratio of the frictional contacts is $\tau_{xz}^{f}/\sigma_{zz}^{f} = \sin\phi$.

The following force balance must be satisfied in steady state

$$\begin{aligned} \sigma_{zz} &= \rho g(h - z)\cos\theta = \sigma_{zz}^{f} + \sigma_{zz}^{c} = \sigma_{zz}^{f} + \frac{k_v}{\tan\alpha}\left(\frac{\mathrm{d}v_x}{\mathrm{d}z}\right)^2 \\ \tau_{xz} &= \rho g(h - z)\sin\theta = \tau_{xz}^{f} + \tau_{xz}^{c} = \sigma_{zz}^{f}\sin\phi + k_v\left(\frac{\mathrm{d}v_x}{\mathrm{d}z}\right)^2 \end{aligned} \tag{19}$$

The stress consists of the contributions from frictional contacts and collisions, described by the Mohr-Coulomb model and the Bagnold-type rheology, respectively. Based on the force balance equations, the following differential equation is derived

$$(1 - \sin\phi/\tan\alpha)k_v \left(\frac{\mathrm{d}v_x}{\mathrm{d}z}\right)^2 - \rho g(\sin\theta - \cos\theta\sin\phi)(h - z) = 0 \tag{20}$$

Solving the above equation, we have the velocity profile

$$v_x(z) = \frac{2}{3}\sqrt{\frac{\rho g(\sin\theta - \cos\theta\sin\phi)}{(1 - \sin\phi/\tan\alpha)k_v}}(h^{3/2} - (h - z)^{3/2}) \tag{21}$$

The velocity profile, evolving as $(h - z)^{3/2}$, corresponds to the Bagnold profile. Once the velocity is obtained, the stress components from frictional contacts and collisions can be determined. From the velocity profile it is noted that velocity and shear rate increase with the inclination θ. With increasing θ, the normal and shear stress induced by collisions become more significant. This increase in the dispersive normal stress results in a trend of dilation, tending to reduce the stress components caused by frictional contacts.

If the granular material is modeled using the $\mu(I)$ model, dense granular flow can only be possible if the inclination θ is between $\tan^{-1}\mu_1$ and $\tan^{-1}\mu_2$. With the unified model, a similar inclination range can be derived as follows

$$\tan\theta = \frac{\tau_{xz}}{\sigma_{zz}} = \frac{(\sigma_{zz} - \sigma_{zz}^c)\sin\phi + \sigma_{zz}^c\tan\alpha}{\sigma_{zz}} = \sin\phi + \frac{\sigma_{zz}^c}{\sigma_{zz}}(\tan\alpha - \sin\phi) \tag{22}$$

The steady dense flow condition requires $0 < \sigma_{zz}^c < \sigma_{zz}$, because if $\sigma_{zz}^c > \sigma_{zz}$ the network of frictional contacts completely disappears, which means that the granular material enters the gaseous regime. Consequently, according to the unified model, the inclination range for dense granular flow is

$$\tan^{-1}(\sin\phi) < \theta < \alpha \tag{23}$$

which is exactly the inclination range derived from the $\mu(I)$ model.

The velocity profiles of the granular flow with two different inclinations are calculated. The material parameters used in the calculation are the same as those used in vertical chute flow. For the unified model, the parameters are taken as $\sin\phi = \mu_1 = 0.38$ and $\tan\alpha = \mu_2 = 0.64$. Regarding how to evaluate parameter k_v, Bagnold [6] proposed a relation based on his experiments on granular-water mixture. However, if the parameter k_v is calculated using Bagnold's relation, the velocity profiles obtained using the unified model and the $\mu(I)$ model have significant discrepancy. Therefore, for dry dense granular flows, a different method for determining k_v is required. Figure 3 shows the results obtained using $k_v = 0.3$ for $\theta = 23°$, and $k_v = 0.1$ for $\theta = 25°$. Using the given parameters, the two models provide similar predictions.

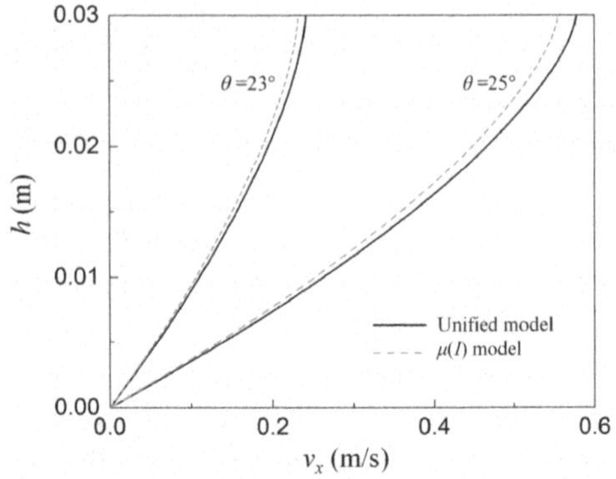

Fig. 3 The velocity profiles obtained using the unified model and the $\mu(I)$ model in the flow over inclined plate

4 Conclusion

A unified constitutive model for granular materials in the quasi-static and dense flow regimes is presented. The model separately and additively considers stress contributions from the two dominate interactions, frictional contact and collision, using the Mohr-Coulomb model and Bagnold-type relation, respectively.

Analytical solutions for granular flows in three configurations, i.e., plane shear, vertical chute flow and flow on inclined plate, are derived based on the presented model. The solutions are compared with those obtained using the $\mu(I)$ model. The following observations can be made. (1) Both models show that in order to switch from the quasi-static to the gaseous regime, one can either increase the shear rate or decrease the pressure. (2) The overall phenomenological frictional coefficient is dependent on flow kinematics, material properties and boundary conditions. In both models there are lower and upper limits for this phenomenological frictional coefficient.

Nevertheless, the unified model has some distinct features. First, the quasi-static solid-like behaviors of granular materials are fully defined in the rate-independent part of the model. Material properties in the quasi-static regime, such as critical state, plastic flow and nonlinear elasticity, can be modeled by changing the quasi-static part to more sophisticated models. Thus, the model is quite flexible. Second, granular materials in the quasi-static and dense flow regime can be described using the model uniformly, which simplifies numerical implementation. Third, the model reveals that in the dense flow regime, especially when granular materials are quite close to the gaseous regime, particle collisions contribute to a significant part of the stress. The $\mu(I)$ model shows that the friction angle increases as the granular material approaches the gaseous regime, but does not give a physical background. The unified model shows that the increase in the phenomenological frictional coefficient is because of particle collisions.

By comparing the solutions from the two models, it is found that the constant k_v requires further investigation. Bagnold [6] proposed a method to calculate k_v based on experiments using water-granular mixture, but we found that this method leads to incorrect predictions for dry granular flows. Future research will focus on how to determine the constant k_v.

Acknowledgements The authors acknowledge the funding from the European Commission Seventh Framework Programme project Multiscale Modelling of Landslides and Debris Flows (MUMOLADE, No. 289911) and HORIZON 2020 project Geohazards Risk Assessment, Mitigation and Prevention (GEO-RAMP, No. 645665). C. Peng wishes to acknowledge the financial support from the Otto Pregl Foundation for Fundamental Geotechnical Research in Vienna.

References

1. Nedderman, R.M.: Statics and Kinematics of Granular Materials. Cambridge University Press, Cambridge (1992)
2. Jop, P., Forterre, Y., Pouliquen, O.: A constitutive law for dense granular flows. Nature **441**(7094), 727–730 (2006)
3. Guo, X.G., Peng, C., Wu, W., Wang, Y.Q.: A hypoplastic constitutive model for debris materials. Acta Geotech. **11**(6), 1217–1229 (2016)
4. Peng, C., Guo, X.G., Wu, W., Wang, Y.Q.: Unified modeling of granular media with smoothed particle hydrodynamics. Acta Geotech. **11**(6), 1231–1247 (2016)
5. Chambon, G., Bouvarel, R., Laigle, D., Naaim, M.: Numerical simulations of granular free-surface flows using smoothed particle hydrodynamics. J NonNewton Fluid Mech. **166**(12), 698–712 (2011)
6. Bagnold, R.A.: Experiments on a gravity-free dispersion of large solid spheres in a Newtonian fluid under shear. R. Soc. Lond. A: Math. Phys. Eng. Sci. **225**, 49–63 (1954)
7. da Cruz, F., Emam, S., Prochnow, M., Roux, J., Chevoir, F.: Rheophysics of dense granular material: discrete simulation of plane shear flows. Phys. Rev. E **72**(2), 021309 (2005)
8. Andreotti, B., Forterre, Y., Pouliquen, O.: Granular media: between fluid and solid. Cambridge University Press, Cambridge (2013)

Measuring the Tensile Strength of *Phleum pratense* L. Roots

Markus Rauchecker, Karin Wriessnig and Wei Wu

Abstract Grasses grow in many environments and are widely used in vegetative techniques to control topsoil erosion. De Baets et al. [6] emphasized the importance of grass roots to increase the topsoil resistance against erosion, especially by concentrated flow. For a better understanding and modeling, the influence of grass roots concerning erosion and soil stability, as well as possible differences in the physical properties of grass species should be known. Referring to reported problems associated with root clamping [15], the aims of this study were (1) to find an appropriate clamping procedure for very fine root measurements practically without damaging the tested roots, and (2) to evaluate the tensile strength, elongation and elasticity of the roots of *Phleum pratense* ("Timothy").

1 Introduction

The effects of vegetation on the stability of slopes and the significant reduction of water erosion rates became highly demanded for protection and solving slope failure and have been reported in many investigations [2, 5, 6, 30, 31, 44, 53, 56, 58, 72]. Also low cost and ecological benefit make the usage of herbaceous or woody plants, or a combination, very advisable.

M. Rauchecker (✉) · W. Wu
Institute of Geotechnical Engineering, University of Natural Resources and Life Sciences, Vienna, Austria
e-mail: markus.rauchecker@boku.ac.at

W. Wu
e-mail: wei.wu@boku.ac.at

K. Wriessnig
Institute of Applied Geology, University of Natural Resources and Life Sciences, Vienna, Austria
e-mail: karin.wriessnig@boku.ac.at

W. Wu (ed.), *Recent Advances in Geotechnical Research*,
Springer Series in Geomechanics and Geoengineering,
https://doi.org/10.1007/978-3-319-89671-7_14

Plants decrease the erodibility of soils by intercepting raindrops and providing additional surface roughness. Transpiration reduces the water content of the soil, thus reducing the pore water pressure. Roots leave macropores which enhance the infiltration capacity, and especially grass roots produce a large biomass with exudates which increase the aggregate stability [4, 31].

Mechanical root reinforcement increases the shear strength of soils and improves plant anchorage [20, 26, 65]. It is influenced by plant species, season, root biomass and distribution in the soil, root morphology, interaction with soil particles and tensile strength of the single root [3, 8, 28, 29, 46].

The root biomass depends on the soil depth, the date within the growing season [45] and the climate.

Generally the tensile strength of roots (Tr) decreases with increasing diameter [10, 12, 42, 44, 47, 50, 65, 69] and can be described by a power law equation [9, 8, 48]. Commandeur and Pyles [18] and Genet et al. [26] found a higher cellulose content per dry mass in roots with a smaller diameter. The cellulose chains are grouped in a hemicellulose matrix, forming microfibrills, and cellulose is responsible for the tensile strength in wood due to its microfibrillar structure [26]. "The structure of cellulose has been found to be optimal for resisting failure in tension" [59] in: [26].

Many authors studied the tensile strength of tree and shrub roots [1, 8, 23, 35, 36, 54, 57, 61, 65, 66], but only few tested grasses [63]. One advantage of grasses is quick germination, so they provide a full plant cover after a short time [11]. Grasses are adapted to resist grazing pressure, thus their tensile strength is relatively high [21].

Though the tensile strength of living roots is higher, to a certain extent even dead roots improve the soil stability for some time [40, 52, 55, 72].

For modelling slope stability the tensile strength of the utilized grass species has to be known. In many studies on grasses Lolium species where used. We tested the perennial grass *Phleum pratense* ("Timothy"), because beneath its agricultural importance, this species is often used in a mixture with legumes and/or other grasses for soil stabilization and erosion control, likewise for cover, filter strips, waterways, and application in other critical areas [49, 60]. *P. pratense* has a shallow root system, tolerates poorly drained soils, even short periods of waterlogging and cold winter temperatures [34, 49].

2 Materials and Methods

In April 2016 *P. pratense* (Timothy), planted in spring 2012 in an agricultural site in Vienna, Austria, was excavated from the soil. The roots were cleaned with tap water and stored at a temperature of +6 °C. To preserve the moisture content the roots were stored in a plastic bag. Single roots were taken just before testing. There was no need for preservation in alcohol, as there was no time for withering.

The root diameters were measured using a microscope (ZEISS 475052-9001) in combination with the software Leica Application Suite V 4.0.0, Build: 877. The

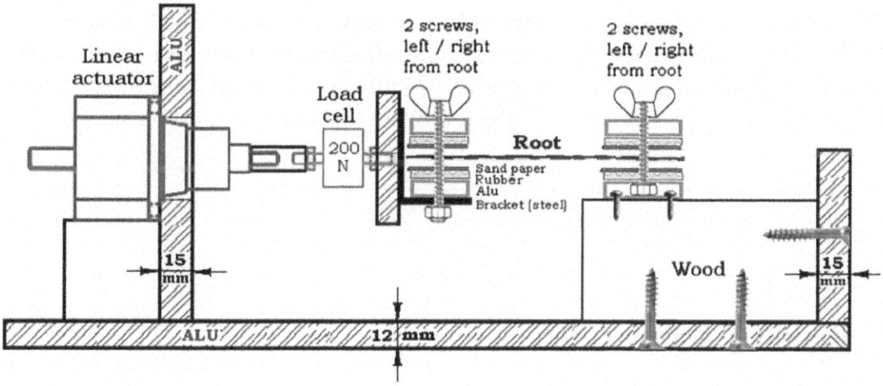

Fig. 1 Root testing machine

diameter was determined to 0.01 mm accuracy at three points. For calculation the lowest diameter was used because the rupture will most likely happen at this point. After measuring, the single root was wrapped in moist paper to avoid drying. The tensile test was performed as soon as possible.

The roots were fixed between the upper and lower part of one clamp (Fig. 1). Both sides of the clamps were covered with ripped rubber, fixed with glue. To increase friction, between the root and the rubber there was a small piece of sand paper K180. The sand paper was folded once, in order that one rough side of the sand paper was always on top and bottom, and could transfer friction to the rubber and the root. The tested root was placed between the two clamps and screwed with caution to prevent rupture at the position of clamping. The distance between the two clamps was 3.2 cm (visible length of the root before test).

To stretch the roots, a Haydon Hybrid Linear Actuator (57000 series) was used. One step, a rotation of 1.8°, is equal to a dislocation of 0.0105 mm, or 10.5 μm. The bipolar, captive, 12 V DC, with a maximal stroke of 31.8 mm was selected. A maximum thrust of 91 kg is delivered. The required 24 V DC power supply of the driver was provided by two 12 V lead accumulators. The velocity was controlled via PC and Basic program to constant 4 mm min^{-1} [64].

The load cell U9C 200 N by Hottinger Baldwin Messtechnik (HBM), accuracy class 0.2, was mounted between the actuator and the left clamp. The right clamp was fixed to a wood block that was connected to the aluminum frame. Further equipment was the amplifier Spider8 and the data acquisition software Catman V5.5 by HBM. Time and force were recorded with a frequency of 10 Hz and time after test converted to the displacement of the left clamp. The test was stopped when the root failed.

Especially small roots are often not straight, so it is not always possible to fix them in the same length. As a starting point of displacement, we chose the point where the (very low) tensile stress of 0.1 N was measured, since we assumed that then the root was stretched usably.

94 roots were selected from 300, the diameters of which ranged from 0.17 to 0.83 mm. Only roots that failed near the middle between the clamps (±7 mm) were chosen for calculation, so we could nearly eliminate the influence from clamping. To calculate tensile resistance (T_r), the following formula was used [9] in: [8]

$$T_r = \frac{F\text{max}}{\pi\left(\frac{D^2}{4}\right)} \tag{1}$$

3 Results and Discussion

3.1 Clamping

Many methods were applied to find the best method for clamping leaves or roots since tensile measurements were made. A number of authors pointed out, that an unsuitable clamping system reduces the measured strength [14, 15, 26, 24, 33, 38]. Clamps can damage the root structure and cause a rupture at the position of clamping, so that the test has to be considered invalid, hence only specimens which break near the middle between the clamps should be selected for calculations [10, 8, 39].

When measuring roots a wide range of different methods and devices were tried for optimizing the fixation of the specimens in the clamps without slipping and damaging. Already Evans [22] and Martens and Booysen [41] tried to enhance cohesion between specimen and clamps or jaws by surfacing these with rubber, cork or emery paper or a combination. Several other authors used similar devices, for instance cementing the ends of root to wooden blocks with epoxy [35], segments of plastic tubing [52], staggered rows of pins [19], stripping the bark and wrapping coarse sandpaper around the core [30], attaching rubber strips and fine sand paper to the screws [8, 71], gluing the ends of the samples into plastic tubes, wrapped capstans and wedge chucks [3], using commercially available crocodile clamps [32], emery paper and fitted clamp form [27], two small wheels that turning inwards clamping the root between [16, 17], self-clamping jaws with a cork spacer [25, 32] lined binder clips and wooden patches [39].

No optimized standard method for the clamping of roots was to be found, thus the method has to be adjusted to the material and the question. After a number of tests we finally used clamps of aluminium, coated with a combination of ripped rubber and folded fine sand paper K180 (Fig. 1) with a result of about 30% success. These clamps could be adjusted precisely to fix even very small roots with a diameter of even 0.2 mm, but it also depends on the skill for reliable results.

Liu [37 in: 70] reported that the axial dimension of a sample should be 8–10 times more than its lateral dimension. Genet et al. [24–25] found the gauge length to be 15 times longer than the central diameter of the root. In root tensile tests at present Zhang [70] found different gauge lengths (grip-to-grip separation) from 50 to even

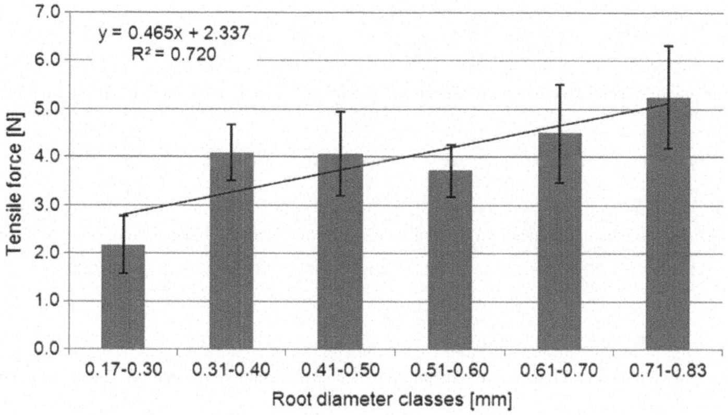

Fig. 2 Tensile force at failure versus root diameter

400 mm. In the present investigation (grass root diameter 0.17–0.83 mm) the gauge length was 3.2 mm ± 1 mm and was found very usable for the chosen root diameters.

3.2 Tensile Strength Tests

A number of tests were first conducted to estimate the clamping force, in order to prevent slippage without destroying the thin grass roots. The results of 94 of 300 root specimens were selected to show the strength characteristics. All values are given as mean ± standard deviation.

The mean tensile force of the roots increases with increasing root diameter (Fig. 2). The diameter of the analyzed roots ranges from 0.17 to 0.83 mm. The minimum and maximum of the recorded values were 0.5 and 8.3 N, respectively. $R^2 = 0.7201$.

To calculate the tensile strength, the tensile force at failure (maximum registered load) was divided by the cross sectional area of the root according to Tosi [64]. Many authors found that the tensile strength of different tree as well as grass roots decreased with increasing root diameter [10, 13, 17, 8, 48].

Burylo et al. [12] tested the effect of different tree, shrub and grass species on soil reinforcement. In their study, the roots of *Achnatherum calamagrostis* L., a perennial grass, had a higher tensile strength than the roots of Quercus pubescens Wild.

De Baets et al. [7] found remarkable differences between different grass species. The tensile strength values published by Comino et al. [17] for *Lolium perenne* are between 365.29 (D = 0.10 mm) and 3, 16 MPa (D = 0.50 mm). The values for *Festuca pratense* range from 372.04 to 6.09 MPa (D = 0.10–0.68 mm). These results are in the same range as our results for *Phleum pratense* L. They point out that the difficulty to measure the diameter of very thin roots might explain the higher variability in small diameter classes.

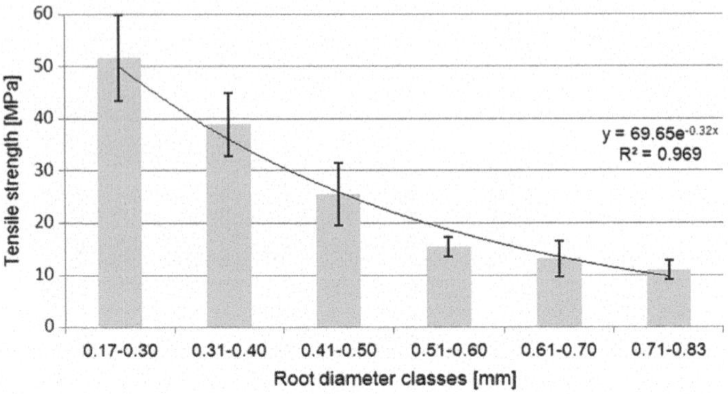

Fig. 3 Tensile strength versus root diameter classes

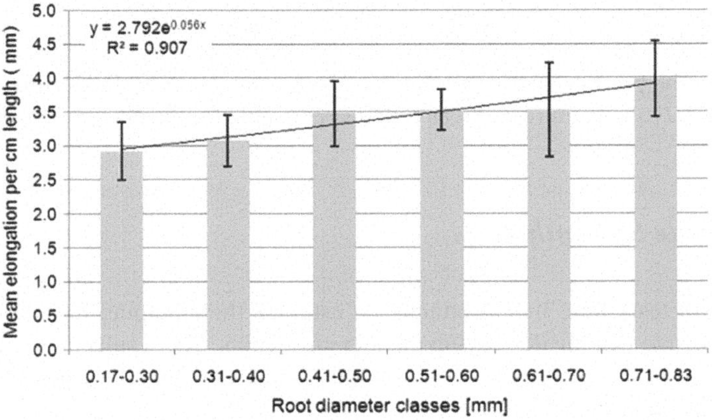

Fig. 4 Mean elongation of the roots per cm versus root diameter class

The mean value of tensile strength of *P. pratense* L. roots decreased significantly with increasing root diameter from 51.7 MPa in the diameter class 0.17–0.30 mm to 11.0 MPa in the diameter class 0.71–0.83 mm (Fig. 3). $R^2 = 0.969$.

The tensile strength is influenced by factors as water content [68], season [15], root architecture [27, 43] or chemical composition [26].

Genet et al. [26] found a significant correlation between the cellulose content and the tensile strength of tree roots. On the other hand, Zhang et al. [69] attributed the tensile strength of pine roots to the higher lignin content and lignin/cellulose ratio in thinner roots. According to Hathaway and Penny [35], the root elasticity is negatively correlated with the lignin/cellulose ratio. Tree roots might be more brittle and weaker in tension than grass roots due to their higher lignin content.

The mean elongation per cm length of *P. pratense* roots increases significantly with increasing root diameter (Fig. 4) from 2.93 mm in the diameter class 0.17–0.30 to 3.98 mm in the diameter class 0.71–0.83 mm. $R^2 = 0.907$.

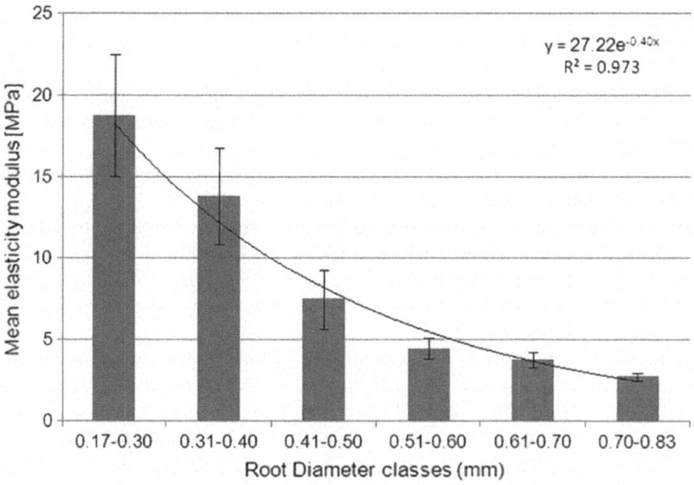

Fig. 5 Modulus of elasticity versus root diameter class

The mean elasticity modulus (Young's modulus, E) decreases from 18.7 MPa (finest roots, average 0.24 mm) to 2.7 MPa (strongest roots, average 0.77 mm) (Fig. 5) $R^2 = 0.973$. The decrease of Young's modulus with increasing diameter was also observed on the roots of rosmary, alfalfa and Cistus plants by Operstein and Frydman [50].

4 Conclusions

The aim of this study was to determine the tensile strength of *P. pratense* L. roots and to optimize the clamping procedure for grass roots. *P. pratense* L. is a component of many seed mixtures used for biological engineering due to its wide ecological amplitude.

The clamping method we used was satisfying, 94 of 300 tested roots were acceptable (for comparison Comino et al. [17]: *Lolium perenne* 35 of 100, *Festuca pratensis* 53 of 100). The success increased during the tests.

Most models for soil reinforcement include tensile strength and root density (root area ratio, RAR) [67], but other parameters as soil texture, root architecture, chemistry (lignin/cellulose) and interaction of root and soil should be included.

The results of our tests are comparable with the data published by Comino et al. [17]. Other grass species will be tested in the near future in order to find out possible differences in tensile strength to other species.

References

1. Abdi, E., Majnounian, B., Rahimi, H., Zobeiri, M.: Distribution and tensile strength of Hornbeam (Carpinusbetulus) roots growing on slopes of Caspian Forests. Iran J. For. Res. **20**(2), 105–110 (2009)
2. Abdullah, M.N., Osman, N., Ali, F.H.: Soil-root shear strength properties of some slope plants. Sains Malaysiana **40**(10), 1065–1073 (2011)
3. Abernethy, B., Rutherfurd, I.D.: The distribution and strength of riparian tree roots in relation to riverbank reinforcement. Hydrol. Process. **15**(1), 63–79 (2001)
4. Amezketa, E.: Soil aggregate stability: a review. J. Sustain. Agric. **14**(2–3), 83–151 (1999)
5. Auerswald, K.: Water erosion. In: Chesworth, W. (ed.) Encyclopedia of Soil Science, pp. 817–822. Springer (2008)
6. De Baets, S., Poesen, J., Gyssels, G., Knapen, A.: Effects of grass roots on the erodibility of topsoils during concentrated flow. Geomorphology **76**, 54–67 (2006)
7. De Baets, S., Poesen, J., Meersmans, J., Serlet, L.: Cover crops and their erosion-reducing effects during concentrated flow erosion. Catena **85**(3), 237–244 (2011)
8. De Baets, S., Poesen, J., Reubens, B., Wemans, K., De Baerdemaeker, J., Muys, B.: Root tensile strength and root distribution of typical Mediterranean plant species and their contribution to soil shear strength. Plant Soil **305**, 207–226 (2008)
9. Bischetti, G.B., Bonfanti, F., Greppi, M.: Misura della resistenza alla trazione delle radici: apparato sperimentale e metodologia d'analisi. Quaderni di Idronomia Montana **21**(1), 349–360 (2003)
10. Bischetti, G.B., Chiaradia, E.A., Simonato, T., Speziali, B., Vitali, B., Vullo, P., Zocco, A.: Root strength and root area ratio of forest species in Lombardy (Northern Italy). Plant Soil **278**(1–2), 11–22 (2005)
11. Brindle, F.: Use of native vegetation and biostimulants for controlling soil erosion on steep terrain. Transp. Res. Rec. J. Transp. Res. Board **1819**, 203–209 (2003)
12. Burylo, M., Rey, F., Mathys, N., Dutoit, T.: Plant root traits affecting the resistance of soils to concentrated flow erosion. Earth Surf. Proc. Land. **37**(14), 1463–1470 (2012)
13. Cazzuffi, D., Corneo, A., Crippa, E.: Slope stabilisation by perennial "gramineae" in southern Italy: plant growth and temporal performance. Geotech. Geol. Eng. **24**(3), 429–447 (2006)
14. Cofie, P., Koolen, A.J.: Test speed and other factors affecting the measurements of tree root properties used in soil reinforcement models. Soil Tillage Res. **63**(1), 51–56 (2001)
15. Cofie, P., Koolen, A.J., Perdok, U.D.: Measurement of stress–strain relationship of beech roots and calculation of the reinforcement effect of tree roots in soil–wheel systems. Soil and Tillage Res. **57**(1), 1–12 (2000)
16. Comino, E., Marengo, P.: Root tensile strength of three shrub species: Rosa canina, Cotoneaster dammeri and Juniperus horizontalis: soil reinforcement estimation by laboratory tests. Catena **82**(3), 227–235 (2010)
17. Comino, E., Marengo, P., Rolli, V.: Root reinforcement effect of different grass species: a comparison between experimental and models results. Soil and Tillage Res. **110**(1), 60–68 (2010)
18. Commandeur, P.R., Pyles, M.R.: Modulus of elasticity and tensile strength of Douglas-fir roots. Can. J. For. Res. **21**(1), 48–52 (1991)
19. Donovan, L.S., Jui, P., Kloek, M., Nicholls, C.F.: An improved method of measuring root strength in corn (Zea mays L.). Can. J. Plant Sci. **62**(1), 223–227 (1982)
20. Dupuy, L., Fourcaud, T., Lac, P., Stokes, A.: Modelling the influence of morphological and mechanical properties on the anchorage of root systems. In: Proceedings of International Conference on 'Wind Effects on Trees' September, pp. 16–18 (2003)
21. Ennos, A.R., Fitter, A.H.: Comparative functional morphology of the anchorage systems of annual dicots. Funct. Ecol. **6**, 71–78 (1992)
22. Evans, P.S.: A study of leaf strength in four ryegrass varieties. N. Z. J. Agric. Res. **7**(4), 508–513 (1964)

23. Fan, C.-C., Su, C.-F.: Effect of soil moisture content on the deformation behaviour of root-reinforced soils subjected to shear. Plant Soil **324**, 57–69 (2009)
24. Genet, M., Kokutse, N., Stokes, A., Fourcaud, T., Cai, X., Ji, J., Mickovski, S.: Root reinforcement in plantations of Cryptomeria japonica D. Don: effect of tree age and stand structure on slope stability. For. Ecol. Manage. **256**(8), 1517–1526 (2008)
25. Genet, M., Stokes, A., Fourcaud, T., Norris, J.E.: The influence of plant diversity on slope stability in a moist evergreen deciduous forest. Ecol. Eng. **36**(3), 265–275 (2010)
26. Genet, M., Stokes, A., Salin, F., Mickovski, S.B., Fourcaud, T., Dumail, J.F., van Beek, R.: The influence of cellulose content on tensile strength in tree roots. Plant Soil **278**(1–2), 1–9 (2005)
27. Ghestem, M., Veylon, G., Bernard, A., Vanel, Q., Stokes, A.: Influence of plant root system morphology and architectural traits on soil shear resistance. Plant Soil **377**(1–2), 43–61 (2014)
28. Ghestem, M., Stokes, A., Cao, K., Kokutse, N.: A new approach for slope stabilization by plant roots on degradation hotspots in Southern China, (2010). www.interpraevent.at/palm-cms/upload_files/Publikationen/Tagungsbeitraege/2010__402.pdf
29. Gray, D.H., Sotir, R.B.: Biotechnical and soil bioengineering slope stabilization: a practical guide for erosion control. Wiley, Hoboken (1996)
30. Greenwood, J.R., Norris, J.E., Wint, J.: Assessing the contribution of vegetation to slope stability. Geotech. Eng. **157**(GE4), 199–207 (2004)
31. Gyssels, G., Poesen, J., Bochet, E., Li, Y.: Impact of plant roots on the resistance of soils to erosion by water: a review. Prog. Phys. Geogr. **29**(2), 189–217 (2005)
32. Hales, T.C., Cole-Hawthorne, C., Lovell, L., Evans, S.L.: Assessing the accuracy of simple field based root strength measurements. Plant Soil **372**(1–2), 553–565 (2013)
33. Hales, T.C., Ford, C.R., Hwang, T., Vose, J.M., Band, L.E.: Topographic and ecologic controls on root reinforcement. J. Geophy. Res. Earth Surf. **114**(F3) (2009)
34. Hall E.: University of Melbourne—pasture species database—Timothy (Phleum pratense) (2008). http://keys.lucidcentral.org/keys/v3/pastures/Html/Timothy.htm
35. Hathaway, R.L., Penny, D.: Root strength in some Populus and Salix clones. NZ J. Bot. **13**(3), 333–344 (1975)
36. Lateh H., Abu Bakar1 M., Khan Y.A., Abustan I. (2011): Influence of tensile force of agave and tea plants roots on experimental prototype slopes. Int. J. Phys. Sci. **6**(18), pp. 4435–4440
37. Liu J.: The effect of dynamic characteristics of beech and larch root strength on trafficability. In: Proceedings of the sixth European ISTVS conference, Vienna, Austria, Sept. 28–30, pp. 1–12 (1994)
38. Loades, K.W., Bengough, A.G., Bransby, M.F., Hallett, P.D.: Biomechanics of nodal, seminal and lateral roots of barley: effects of diameter, waterlogging and mechanical impedance. Plant Soil **370**(1–2), 407–418 (2013)
39. Manzur, M.E., Hall, A.J., Chimenti, C.A.: Root lodging tolerance in Helianthus annuus (L.): associations with morphological and mechanical attributes of roots. Plant Soil **381**(1–2), 71–83 (2014)
40. Marcandella, D., Rickli, Ch. (2010). Zugfestigkeit von Wurzeln abgestorbener Bäume. FAN-Agenda 2/10. *FAN Fachleute Naturgefahren Schweiz (Eds.)*
41. Martens, P.O., de V. Booysen, P.: A tensilmeter for the measurement of the tensile strength of grass leaf blades 1. Proc. Annu. Congresses Grassland Soc. South. Afr. **3**(1):51–56 (1968)
42. Mattia, C., Bischetti, G.B., Gentile, F.: Biotechnical characteristics of root systems of typical Mediterranean species. Plant Soil **278**(1–2), 23–32 (2005)
43. Mickovski, S.B., Van Beek, L.P.H.: Root morphology and effects on soil reinforcement and slope stability of young vetiver (Vetiveriazizanioides) plants grown in semi-arid climate. Plant Soil **324**(1–2), 43–56 (2009)
44. Naghdi, R., Maleki, S., Abdi, E., Mousavi, R., Nikooy, M.: Assessing the effect of Alnus roots on hillslope stability in order to use in soil bioengineering. J. For. Sci. **59**(11), 417–423 (2013)
45. Niemi, R.M., Vepsäläinen, M., Wallenius, K., Simpanen, S., Alakukku, L., Pietola, L.: Temporal and soil depth-related variation in soil enzyme activities and in root growth of red clover (Trifoliumpratense) and timothy (Phleum pratense) in the field. Appl. Soil. Ecol. **30**(2), 113–125 (2005)

46. Nilaweera, N.S., Nutalaya, P.: Role of tree roots in slope stabilisation. Bull. Eng. Geol. Environ. **57**(4), 337–342 (1999)
47. Norris, J.E.: Root reinforcement by hawthorn and oak roots on a highway cut-slope in Southern England. In: Eco-and Ground Bio-Engineering: The Use of Vegetation to Improve Slope Stability, pp. 61–71. Springer, Netherlands (2007)
48. Nyambane, O.S., Mwea, S.K.: Root tensile strength of 3 typical plant species and their contribution to soil shear strength; a case study: SasumuaBackslope, Nyandarua District, Kenya. J. Civil Eng. Res. Pract. **8**(1), 57–73 (2011)
49. Ogle, D.G., St. John, L., Tilley, D.J.: Plant Guide for Timothy (Phleum pratense). USDA-Natural Resources Conservation Service, Idaho State Office. Boise, ID (2011)
50. Operstein, V., Frydman, S.: The influence of vegetation on soil strength. Proc. Inst. Civil Eng.-Ground Improv. **4**(2), 81–89 (2000)
51. O'Loughlin, C.L., Watson, A.L.E.X.: Root-wood strength deterioration in radiata pine after clearfelling. NZJ For. Sci. **9**(3), 284–293 (1979)
52. O'Loughlin, C., Ziemer, R.R.: The importance of root strength and deterioration rates upon edaphic stability in steepland forests. Proceedings of I.U.F.R.O. Workshop P.1.07–00 Ecology of Subalpine Ecosystems as a Key to Management. 2–3 Aug 1982, Corvallis, Oregon. Oregon State University, Corvallis, Oregon, pp. 70–78 (1982)
53. Petrone, A., Preti, F.: Soil bioengineering for risk mitigation and environmental restoration in a humid tropical area. Hydrol. Earth Syst. Sci. **14**(2), 239–250 (2010)
54. Pohl, M., Stroude, R., Körner, C., Buttler, A., Rixen, C. (2009, April). Root diversity in alpine plants: root length, tensile strength and plant age. In: EGU General Assembly Conference Abstracts, vol. 11, p. 11089
55. Preti, F.: Forest protection and protection forest: tree root degradation over hydrological shallow landslides triggering. Ecol. Eng. **61**, 633–645 (2013)
56. Preti, F., Giadrossich, F.: Root reinforcement and slope bioengineering stabilization by Spanish Broom (Spartiumjunceum L.). Hydrol. Earth Syst. Sci. **13**(9), 1713–1726 (2009)
57. Schwarz, M., Phillips, C., Marden, M., McIvor, I.R., Douglas, G.B., Watson, A.: Modelling of root reinforcement and erosion control by 'Veronese'poplar on pastoral hill country in New Zealand. NZ J. Forest. Sci. **46**(1), 1 (2016)
58. Schwarz, M., Preti, F., Giadrossich, F., Lehmann, P., Or, D.: Quantifying the role of vegetation in slope stability: a case study in Tuscany (Italy). Ecol. Eng. **36**(3), 285–291 (2010)
59. Sjostrom, E.: Wood Chemistry Fundamentals and Applications, 2nd edn, p. 293. Academic Press Inc, San Diego (1993)
60. Šlezingr, M., Foltýnová, L., Šulc, V.: The design procedure for pre-grown stabilization reinforced grass carpet. Acta Universitatis Agriculturae et Silviculturae Mendelianae Brunensis **59**(6), 355–358 (2014)
61. Stokes, A., Atger, C., Bengough, A.G., Fourcaud, T., Sidle, R.C.: Desirable plant root traits for protecting natural and engineered slopes against landslides. Plant Soil **324**(1–2), 1–30 (2009)
62. Styczen, M.E., Morgan, R.P.C.: Engineering properties of vegetation. In: Morgan, R.P.C., Rickson, R.J. (eds.) Slope Stabilization and Erosion Control: A Bioengineering Approach, pp. 5–58. E & F.N, Spon, London (1995)
63. Tengbeh, G.T.: The effect of grass roots on shear strength variations with moisture content. Soil Technol. **6**(3), 287–295 (1993)
64. Tosi, M.: Root tensile strength relationships and their slope stability implications of three shrub species in the Northern Apennines (Italy). Geomorphology **87**(4), 268–283 (2007)
65. Trükmann, K., Horn, R.: Stabilisierungseffekte von Pflanzenwurzeln als Möglichkeit zur Reduzierung der mechanischen Bodendeformationen in Grünland. Tagungsbeitrag zu: Jahrestagung der DBG, Kom. VIII. Titel der Tagung: Böden—eine endliche Ressource. Veranstalter: DBG, Sept 2009, Bonn. Berichte der DBG (nicht begutachtete online Publikation) (2009). http://www.dbges.de
66. Vergani, C., Chiaradia, E.A., Bischetti, G.B.: Variability in the tensile resistance of roots in Alpine forest tree species. Ecol. Eng. **46**, 43–56 (2012)

67. Wu, T.H., McKinnell III, W.P., Swanston, D.N.: Strength of tree roots and landslides on Prince of Wales Island, Alaska. Can. Geotech. J. **16**(1), 19–33 (1979)
68. Yang, Y., Chen, L., Li, N., Zhang, Q.: Effect of root moisture content and diameter on root tensile properties. PLoS ONE **11**(3), e0151791 (2016). https://doi.org/10.1371/journal.pone.0151791
69. Zhang, C.B., Chen, L.H., Jiang, J.: Why fine tree roots are stronger than thicker roots: the role of cellulose and lignin in relation to slope stability. Geomorphology **206**, 196–202 (2014)
70. Zhang, C., Chen, L., Jiang, J., Zhou, S.: Effects of gauge length and strain rate on the tensile strength of tree roots. Trees **26**(5), 1577–1584 (2012)
71. Zhong, R.H., He, X.B., Bao, Y.H., Tang, Q., Gao, J.Z., Yan, D.D., … Li, Y.: Estimation of soil reinforcement by the roots of four post-dam prevailing grass species in the riparian zone of Three Gorges Reservoir, China. J. Mt. Sci. **13**(3):508–521 (2016)
72. Ziemer, R.R. (1981). Roots and the stability of forested slopes. Publication No. 132. Int. Assoc. Hydrol. Sci. 343–361

Shallow Landslides: Retrospective Analysis of the Protective Effects of Forest and Conclusions for Prediction

Christian Rickli, Peter Bebi, Frank Graf and Christine Moos

Abstract Landslides repeatedly cause damage in steep terrain and, therefore, the prediction of landslide susceptibility is important. In this context, comprehensive inventories with detailed information on shallow rainfall triggered landslides have been established in Switzerland. The present database includes information on more than 600 individual landslides. The data illustrate that slope inclination belongs to the most decisive parameters. Landslides occurred on slopes with inclinations from 20° to 50° with a majority on slopes of 35°–40°. The angles of internal friction of the landslide soil mostly range between 31° and 35°. Vegetation effects, particularly the existence of forest, are important reasons that slopes with inclinations of more than 35° were stable before the respective hazard events. This is supported by the fact that, in the majority of the event-inventories, landslide density is lower in forested areas than on open land. However, not only the presence of forest has an effect on slope stability but also its condition. Additionally, terrain morphology near the landslide seems to be relevant. A filtering procedure based on (i) soil mechanics and slope inclination, (ii) forest structure and (iii) terrain morphology explained more than 95% of 218 landslides in forested terrain. Additional case studies confirmed that forest structural characteristics do have an important influence on root reinforcement and shallow landslide susceptibility. For example a gap length of >20 m seems to be

C. Rickli (✉)
Swiss Federal Institute for Forest, Snow and Landscape Research WSL, Birmensdorf, Switzerland
e-mail: christian.rickli@wsl.ch

P. Bebi · F. Graf · C. Moos
WSL Institute for Snow and Avalanche Research SLF, Davos Dorf, Switzerland
e-mail: bebi@slf.ch

F. Graf
e-mail: graf@slf.ch

C. Moos
e-mail: christine.moos@bfh.ch

C. Moos
School of Agricultural, Forest and Food Science HAFL, Berne University of Applied Sciences, Zollikofen, Switzerland

W. Wu (ed.), *Recent Advances in Geotechnical Research*,
Springer Series in Geomechanics and Geoengineering,
https://doi.org/10.1007/978-3-319-89671-7_15

Fig. 1 **a** A shallow landslide near St. Antönien (no. 6 on the map). **b** Location of Swiss landslide inventories (cf. Table 1)

a critical threshold for landslide triggering. The study confirms that forest structure has—together with geomorphological and hydrological conditions—an important influence on shallow landslide susceptibility.

1 Introduction

Shallow landslides triggered by intensive rainfall are common hazards in alpine regions. They pose considerable danger to humans and infrastructure, particularly because they often evolve into hillslope debris flows with long runout and due to the difficulty of predicting/forecasting their spatial and temporal occurrence. The total damage sum due to landslides from 1972 to 2007 in Switzerland was estimated to 520 million Euros [7]. In 2005, one large devastating event (in Switzerland) triggered more than 5000 shallow landslides within only three days [2]. Although shallow landslides commonly occur both on areas with forest cover (Fig. 1) and on open land, the positive effect of forest due to tree roots as well as enhanced interception and evaporation on slope stability is well recognised [9, 18].

Landslide-event inventories contribute to a better understanding of the hazard processes and are, therefore, an important basis for hazard assessment and mitigation

Table 1 Landslide-event inventories by the WSL in Switzerland. More than 60 parameters including volume, inclination, and soil material were recorded per landslide

Location	No. of inventory*	Date of event	Precipitation		No. of landslides documented	Area of watershed (km^2)
			Cum. rainfall (mm)	Duration (h)		
Sachseln	1	15.8.1997	150	2	280	8.2
Appenzell	2	31.8.–1.9.2002	120	9	107	10.2
Napf	3	15.–16.7.2002	60	3	64	2.5
Surselva	4	14.–16.11.2002	252	63	35	3.2
Entlebuch	5	18.–23.8.2005	269	72	66	5.1
St. Antönien	6	18.–23.8.2005	185	72	63	4.7
Napf	7	18.–23.8.2005	241	72	54	1.6
Eriz	8	4.7.2012	60–100	2	25	9.5
Total					694	45.0

*Locations of the inventories are indicated in Fig. 1b

[e.g. 6, 8]. This article combines the analysis of landslide-event inventories with new field measurements in order to investigate relevant triggering factors and in particular the effect of forest structure on landslide susceptibility.

2 Data and Methods

The present study is mainly based on several comprehensive and field-based landslide-event inventories with detailed information on 694 shallow landslides (Table 1). The field surveys were performed with analogue protocols, covering many relevant parameters including the dimensions of the slides, and site characteristics such as vegetation, geomorphology, and topography as well as characteristics of the subsequent runout (see [16] for more details). Relating to soil mechanics, the soil of all landslides was assessed and classified according to VSS standard SN 670 008a [20] in the field. In addition, the correspondent angles of internal friction Φ' were derived following VSS standard SN 670 010 [21]. Forest condition near the landslides was assessed in two different ways: (1) based on guidelines that describe requirements for sustainable protection forests adapted to the site according to Frehner et al. [3] and (2) using a general forest stand code with assessments of forest layering, cover, development and mixture [13]. The underlying database currently includes information from landslide inventories with more than 600 individual landslides [17, Table 1].

The landslide data and some of the preliminary results of the inventories were used to establish an evaluation procedure of landslides triggered in forest areas [5]. This procedure ("filter") intends to retrospectively explain the occurrence of shallow landslides. The subsequent 3 fields were characterised by 6 parameters with the below mentioned values that explain failure:

1. Soil mechanics: slope inclination is more than 5° steeper than the angle of internal friction of the soil [4].
2. Forest characteristics:
   - layering is unstocked or mono-layered: layering according to portion of individuals in upper, medium, and lower layer (i.e. >2/3 of stand height, 1/3–2/3 of stand height, <1/3 of stand height)
   - tree cover is partial (~40%) or sparse (~20%): ratio of crown cover (canopy) projection (without multiple cover) to total cover (at the most 100%)
   - development is unstocked or a young stand up to thicket
   - forest condition: special case event-inventory Sachseln with forest condition "poor" [14] with gaps, signs of terrain movements, old landslides or wind throw, as well as inappropriate species mixture [3].
3. Terrain morphology: curvature line of slope-transverse profile is convex-flat, flat-concave, or convex-concave.

In addition to the aforementioned documentation of the landslides, further analyses on the effect of forest structure on landslide susceptibility were performed in two of the study areas (St. Antönien with 46 and Sachseln with 136 landslides in forested terrain). For each landslide plot as well as for a comparable set of control points (same number and range of forest cover and steepness), variables characterising forest structure, topography, geomorphology and hydrology were measured in the field (St. Antönien) and derived from LiDAR (light detection and ranging, 2 m resolution) based elevation data (St. Antönien and Sachseln) and from a stand classification (Sachseln). Subsequently, the data were analysed with univariate tests (Wilcoxon rank-sum tests for continuous and Chi-square tests for categorical variables) as well as multivariate statistics including logistic regression models and classification trees [11, 12]. For the case study of St. Antönien, the root distribution was additionally measured at regular distances from seven trees to calibrate a root bundle model and to quantify root reinforcement [10]. Based on this model, a proxy-variable for root reinforcement of the nearest tree was calculated [12].

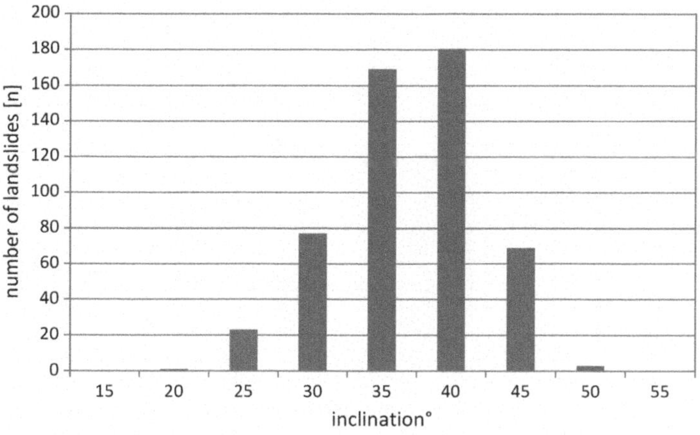

Fig. 2 Slope inclinations at the release areas of the landslides

3 Results and Discussion

3.1 *Landslide Inventories*

According to the landslide database the slope inclination at starting points is a decisive parameter for landslide occurrence, i.e. steep slopes are more prone to hillslope processes. All inventories (Table 1) show that landslide occurrence starts at slope inclinations of 20° and increases with steepening slope angel up to about 35°–40°. Roughly half of all landslides (48%) occurred on slopes with inclinations of more than 35° (Fig. 2).

Almost 80% of the soils were classified as lean clay, silt, clayey to silty gravel, silty gravel and silty sand (Fig. 3). The corresponding angles of internal friction Φ' ranged between 27° and 38° with the most frequent soil types materials found in the interval 31°–35°. In a simplified approach slopes of bare soil can be regarded stable up to inclinations equal to the angle of internal friction of the pure soil material. The reasons that steep terrain with inclinations between 35° and 50° was stable before the landslide triggering rainfall events can be found in parameters such as cohesion, compaction/compression and vegetation effects.

In the majority of the event-inventories, landslide density is lower in forested areas than on open land [15] demonstrating a considerable effect of forest on slope stability. Yet, landslide occurrence not only depends on the presence of forest but also on the forest stand characteristics at the release area [14]. The assessments of the forests near the landslides based on [3] identify high landslide density in forested areas with unfavourable stand characteristics, e.g. species and stand structure not suitable to site, many gaps, parts previously damaged, patchy stands. Vegetation assessment using the forest stand code showed that many landslides occurred in less favourable

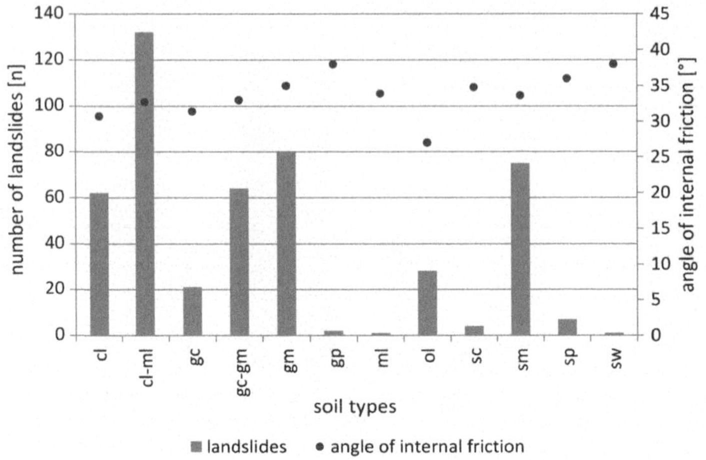

Fig. 3 Number of landslides (blue bars) related to the different soil types, including information for the angle of internal friction Φ' (black dots) according to [21]

stands, e.g. unstocked, mono-layered, young stand/thicket, more than 80% needle trees, and partial (40%) or sparse (20%) tree coverage.

Terrain morphology is broadly considered relevant for landslide occurrence [e.g. 1, 19]. Our data show that a considerable proportion of all landslides (40%) was released in morphology types known to be susceptible to shallow landslides such as edges of terraces, hollows and gullies (curvatures line of slope—transverse profile convex-flat, flat-concave, or convex–concave).

3.2 *Filtering Procedure*

The serially applied filtering procedure of shallow landslides triggered during heavy rainfall events in forested areas, retrospectively explained 50% of the landslides by soil mechanics, 40% by forest characteristics and 7% by terrain morphology [5]. In total 97% of 218 landslides during the heavy rainfall events in 1997 (Sachseln), 2002 (Napf, Appenzell) and 2005 (Entlebuch, Napf, St. Antönien) were explained and only six (3%) of the 218 investigated landslides stayed unresolved. While new remote sensing data provide an increasing potential to deduce variables related to topography and forest structure on higher resolution, the assessment of the angle of internal friction Φ' remains challenging. In order to characterise larger areas, e.g. catchments, in respect of soil mechanics, a conventionally applied approach is to assign fixed values of Φ'—and cohesion c'—to different geological units. However, this is not entirely satisfactory as is demonstrated for the Drusberg layers in Sachseln with 67 landslides. There, the assigned friction angle Φ' is given with 36°. Yet, 95% of the events where triggered in soil with $\Phi' = [27°, 33°]$. Consequently, new approaches

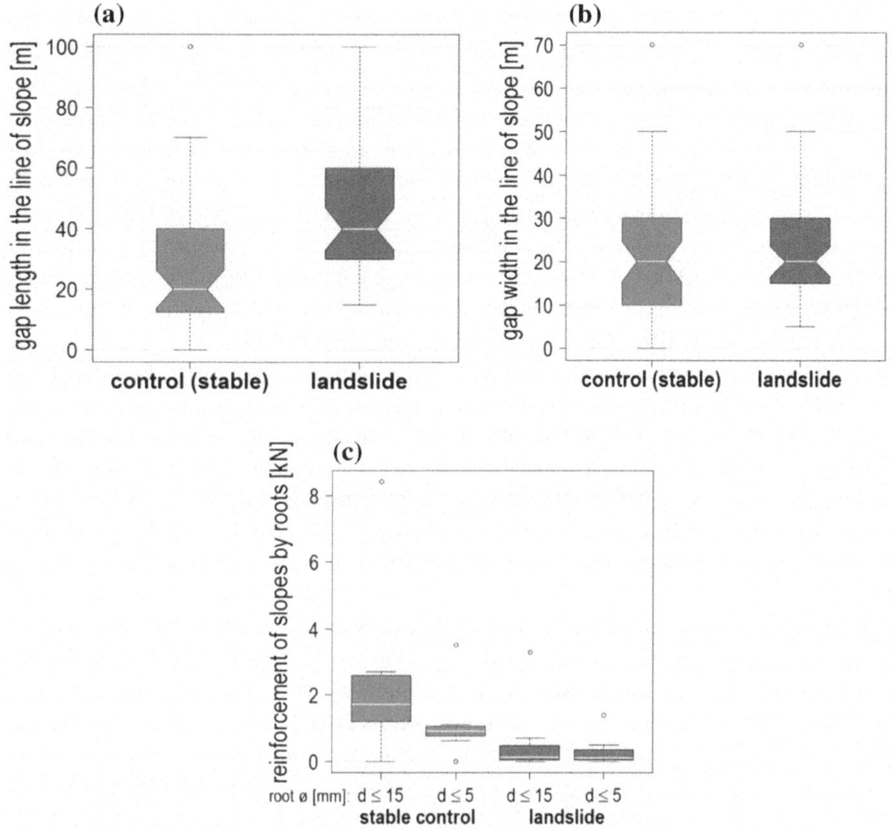

Fig. 4 **a** Distribution of gap lengths (significant) and **b** widths (not significant) of the largest gaps in control and landslide plots in St. Antönien and **c** root reinforcement by tree roots in a subset of plots for roots with a diameter <15 and <5 mm (significant) [12]

will be necessary to improve existing estimations for Φ'—possibly combined with other filtering parameters.

3.3 *Influence of Forest Structure and Other Parameters on Landslide Susceptibility*

In the study area St. Antönien, landslide and control plots were characterized by a relatively open forest structure (15–70% degree of forest cover). However, gap length measured in landslide plots was significantly larger compared to gap length of control plots ($p < 0.001$; Fig. 4a). In contrast, gap width in control and landslide plots did not differ significantly (Fig. 4b).

The distance from the (potential) release point to the nearest tree as well as the mean distance to the five nearest trees measured in the field were significantly higher in landslide plots compared to control plots ($p = 0.002$; $p = 0.001$). A similar result was obtained for the distance measured on the orthophoto of 2003 (before landslide event; $p < 0.001$). Moreover, the distances to the nearest tree on the orthophoto did not significantly differ from the distances measured in the field. The root reinforcement in a subset of landslide and control plots as well as the proxy-variable for root reinforcement of the nearest tree was significantly higher in control plots than in landslide plots ($p < 0.001$; Fig. 4c). Tree coverage calculated from the canopy height model (controlled variable), tree height, successional stage and layering of the forest did not significantly differ between landslide and control plots.

The classification tree model fitted with gap length and slope angle, first partitions the data based on a threshold of 20 m for gap length. In plots where the maximum gap is longer than 20 m, the probability of landslide occurrence is distinctly reduced, if the slope angle α is smaller than 36° (Fig. 5a). The distance to the nearest tree as explanatory variable, partitions the data set with a threshold of 6.2 m. In the case of smaller distances to the nearest tree, the probability of landslides is enhanced, given the slope angle α is larger than 38° (Fig. 5b) or, if there are signs of water logging (Fig. 5c).

In the study area of Sachseln (mixed forest with *Abies alba*, *Picea abies* and *Fagus silvatica*), tree height, standard deviation of tree height were significantly lower compared to the Study area of St. Antönien. Furthermore, stands affected by landslides were younger and less deciduous. Moos et al. [12] found that multivariate models integrating forest structure performed better than models solely based on terrain and hydrological variables, and models based on field variables performed much better than models based on LiDAR derived variables. Trees in Sachseln are on average taller and have bigger diameters at breast height (dbh) in control plots than in landslide plots, indicating enhanced reinforcement of the soil with increasing tree age. As the forest stands were affected by wind-throw and beetle calamity in the 1970s and again in 1990 (storm "Vivian"), we can assume that landslides occurred in a critical post-disturbance phase in which roots of dead trees did not reinforce the soil anymore and roots of the upcoming young growth were still too weak to stabilize the soil.

4 Conclusions

The study confirms that not only the presence of forest but also forest structural characteristics do have an important influence on shallow landslide susceptibility. It further turned out that the presented filtering procedure as well as landslide susceptibility maps may potentially consider additional criteria. In particular, gap length and distance between trees seem to be relevant variables to include in a further developed filter. However, the varying results between the two study sites make also evident that the influence of different forest structural characteristics depends on regional and

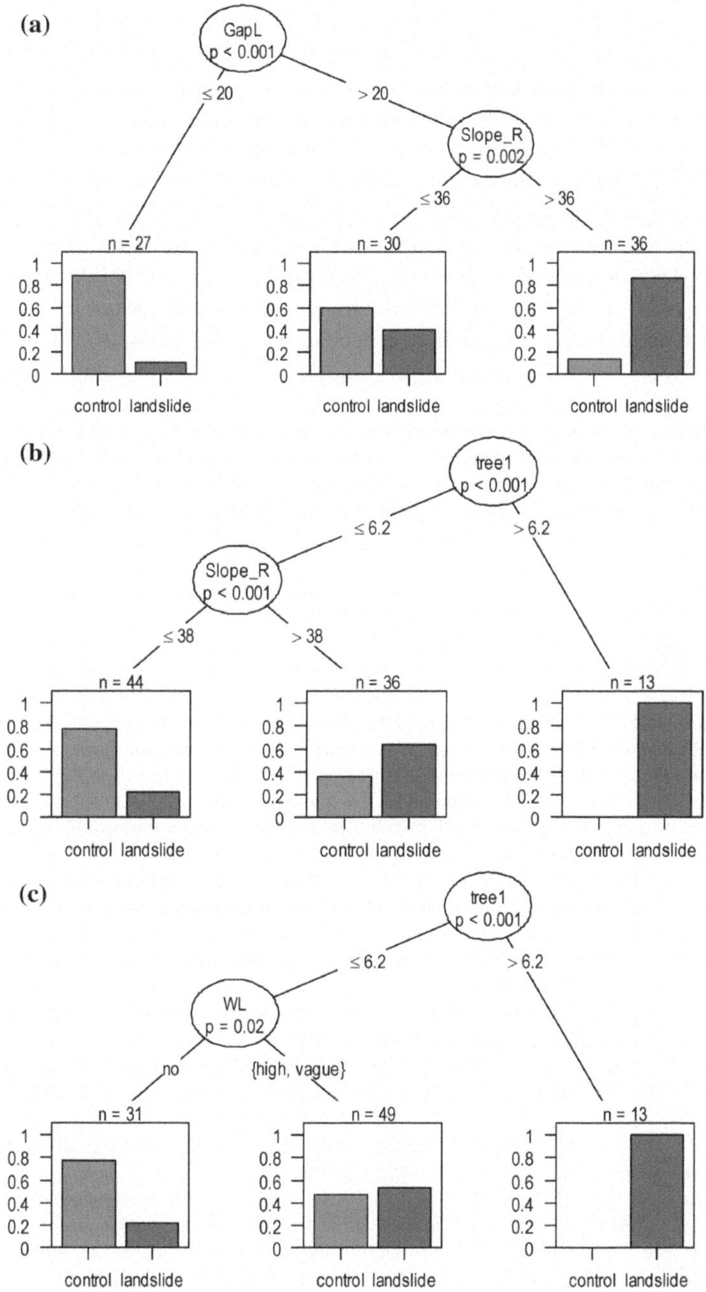

Fig. 5 Classification trees to predict landslide occurrence in St. Antönien based on significant forest structural variables and selected terrain variables. The models were fitted with the whole dataset (n = 93). The n-values exhibit the number of cases explained by the corresponding variables. GapL = length of the largest gap, Slope_R = slope of the release point, tree1 = distance to the nearest tree, WL = signs of water logging [12]

local conditions, such as vegetation zones, altitudinal range, disturbance history and the triggering rainfall event. The findings both confirm and complement the Swiss directives for the management of protection forests [3]. Such management activities include not only the regulation of the tree species composition but also the promotion of canopy openings which are large enough to provide sustainable regeneration on the one hand but are not too long (<20 m) to cause landslides on the other hand. In addition, the results help to quantitatively assess relevant spatial data for slope stability (in particular related to forest structure) and to identify possibilities and limitations for further development of tools for a better prediction of landslide susceptibility. Finally, the study indicates that landslide inventories are important components for improving hazard assessment tools to reduce damage due to shallow landslides and to establish recommendations for appropriate land use and forest management.

Acknowledgements We thank Josias Mattli for his contribution to the quantification of root reinforcement in the study area of St. Antönien. Furthermore, we acknowledge the project SOSTANAH of the Swiss National Research Programme 68 (grant number 406840_143122), and the EU project REVENUES FP7-PEOPLE-2012-IAPP (grant number 324,466) for supporting and funding our research.

References

1. Aleotti, P., Baldelli, P., Polloni, G.: Landsliding and flooding event triggered by heavy rains in the Tanaro basin (Italy). Proceedings of International Symposium Interpraevent 1996, vol. 1, pp. 435–446. Garmisch-Partenkirchen (1996)
2. Bezzola, G.R., Hegg, C.: Ereignisanalyse Hochwasser 2005, Teil 1—Prozesse, Schäden und erste Einordnung. Bundesamt für Umwelt BAFU, Eidg. Forschungsanstalt WSL. Umwelt-Wissen Nr. 0707. P. 215 (2007)
3. Frehner, M., Wasser, B., Schwitter, R.: Nachhaltigkeit und Erfolgskontrolle im Schutzwald. Wegleitung für Pflegemassnahmen in Wäldern mit Schutzfunktion, Vollzug Umwelt, Bundesamt für Umwelt, Wald und Landschaft, Bern, p. 564 (2005)
4. Graf, F., Frei, M., Böll, A.: Effects of vegetation on the angle of internal friction of a moraine. FOS-NOLA **82**, 61–78 (2009)
5. Graf, F., Rickli, C.: Simple explanations for shallow landslides!? Geophysical Research Abstracts, vol. 18. EGU General Assembly (2016)
6. Guzzetti, F., Cesare, A.M., Cardinalia, M., Fioruccia, F., Santangeloa, M., Changc, K.-T.: Landslide inventory maps: New tools for an old problem. Earth Sci. Rev. **112**(1–2), 42–66 (2012)
7. Hilker, N., Badoux, A., Hegg, C.: The Swiss flood and landslide damage database 1972-2007. Nat. Hazards Earth Syst. Sci. **9**(3), 913–925 (2009)
8. Hürlimann, M., McArdell, B.W., Rickli, C.: Field and laboratory analysis of the runout characteristics of hillslope debris flows in Switzerland. Geomorphology **232**, 20–32 (2015)
9. Kim, D., Im, S., Lee, C., Woo, C.: Modeling the contribution of trees to shallow landslide development in a steep, forested watershed. Ecol. Eng. **61**, 658–668 (2013)
10. Mattli, J.: Quantifizierung der Wurzelverstärkung im Schutzwald, Bachelor Thesis. University of Applied Sciences, Berne, p. 79 (2014)
11. Moos, C.: How Does Forest Structure Affect Landslide Susceptibility? Statistical Prediction Models for Shallow Landslides Integrating Forest Structure, Master Thesis. ETH Zurich, p. 97 (2014)

12. Moos, C., Bebi, P., Graf, F., Mattli, J., Rickli, C., Schwarz, M.: How does forest structure affect root reinforcement and susceptibility to shallow landslides? Earth Surf. Proc. Land. (2016). https://doi.org/10.1002/esp.3887
13. Rickli, C. (ed.) (2001) Vegetationswirkungen und Rutschungen. Untersuchung zum Einfluss der Vegetation auf oberfächennahe Rutschprozesse anhand der Unwetterereignisse in Sachseln OW am 15. August 1997. Birmensdorf, Bern, Eidg. Forschungsanstalt WSL, Bundesamt für Umwelt, Wald und Landschaft, p. 97
14. Rickli, C., Zürcher, K., Frey, W., Lüscher, P.: Wirkungen des Waldes auf oberflächennahe Rutschprozesse. Schweiz. Z. Forstwes. **153**(11), 437–445 (2002)
15. Rickli, C., Raetzo, H., McArdell, B., Presler, J.: Hanginstabilitäten.: In: Bezzola, G.R., Hegg, C. (eds.) Ereignisanalyse Hochwasser 2005. Teil 2—Analyse von Prozessen, Massnahmen und Gefahrengrundlagen. Bern, Bundesamt für Umwelt BAFU, Birmensdorf, Eidgenössische Forschungsanstalt WSL, pp. 97–116 (2008)
16. Rickli, C., Graf, F.: Effects of forests on shallow landslides—case studies in Switzerland. For. Snow Landscape Res. **82**(1), 33–44 (2009)
17. Rickli, C., McArdell, B., Badoux, A., Loup, B.: Database shallow landslides and hillslope debris flows. [Abstract] In: Koboltschnig, G. (ed.) 13th Congress Interpraevent 2016. 30 May to 2 June 2016, Lucerne, Switzerland. Extended Abstracts. Living with natural risks, pp. 242–243. Luzern, International Research Society Interpraevent (2016)
18. Schwarz, M., Preti, F., Giadrossich, F., Lehmann, P., Or, D.: Quantifying the role of vegetation in slope stability: a case study in Tuscany (Italy). Ecol. Eng. **36**(3), 285–291 (2010)
19. von Ruette, J., Papritz, A., Lehmann, P., Rickli, C., Or, D.: Spatial statistical modeling of shallow landslides—Validating predictions for different landslide inventories and rainfall events. Geomorphology **133**, 11–22 (2011)
20. VSS: Identifikation der Lockergesteine – Labormethode mit Klassifikation nach USCS. Schweizer Norm SN 670 008a. Vereinigung Schweizerischer Strassenfachleute VSS, p. 16 (1997)
21. VSS: Geotechnische Erkundung und Untersuchung – Geotechnische Kenngrössen. Schweizer Norm SN 670 010. Vereinigung Schweizerischer Strassenfachleute VSS, p. 19 (2011)

Simulation of Rainfall-Induced Landslide of the Vegetated Slope

Barbara Maria Świtała and Wei Wu

Abstract Soil bioengineering methods offer an efficient and eco-friendly solution for slope stabilisation. These measures are especially effective in protection against shallow, rainfall-induced landslides, in which the potential slip surface crosses root zones. Vegetation contributes to the enhancement of the slope stability in several ways, but there are two factors which are the most significant, namely, mechanical root reinforcement and evapotranspiration. The model proposed in this paper allows simulation of rainfall-induced landslides of vegetated slopes, taking into account both factors. The model is tested performing the numerical example, based on the case study. Stability of the bare and vegetated slope is investigated and obtained results are compared and analysed. Rapid accumulation of displacements indicates a moment of failure. Stability enhancement due to the vegetation is defined in reference to the stability of a non-vegetated slope, as the delay of the occurrence of a rainfall-induced landslide.

1 Introduction

Soil bioengineering methods can be effectively used as environmentally-friendly and economically attractive solutions of slope stabilisation. These measures are especially favourable when considering rainfall-induced landslides, which are character-

B. M. Świtała (✉) · W. Wu
Institute of Geotechnical Engineering, Feistmantelstrasse 4, 1180 Vienna, Austria
e-mail: barbara.m.switala@gmail.com

W. Wu
e-mail: wei.wu@boku.ac.at

B. M. Świtała
Institute of Hydro-Engineering, Polish Academy of Sciences,
Kościerska 7, 80-328 Gdańsk, Poland

W. Wu (ed.), *Recent Advances in Geotechnical Research*,
Springer Series in Geomechanics and Geoengineering,
https://doi.org/10.1007/978-3-319-89671-7_16

ized by shallow, translational slip surfaces. If the enhancement of the stabilisation provided by plants is not sufficient, a combined solution with structural measures can also be applied.

There are two main stabilisation factors which result from the presence of vegetation on a slope. The first factor is the mechanical root reinforcement which provides a structural stabilisation. Soil reinforced with plant roots can be compared with the concrete reinforced with steel or fibreglass. Soil, similarly to concrete, is weak in tension and strong in compression, whereas steel and roots can sustain large tension, but are very weak in compression. Consequently, soil-root composite is an optimum, balanced combination which benefits from the strong points of both soil and roots. Roots can be divided into two different kinds: (i) thick, woody roots which behave similarly to soil nails and cross the potential slope surface, or anchor the weak soil layer to the stable bedrock; (ii) thin, hairy roots which contribute to the enhancement of the soil cohesion, binding soil particles.

The second factor is transpiration. Soil moisture is absorbed by thin plant roots and transferred through the xylem of woody roots to the stem and leaves. The driving force for the transpiration process is the gradient of potentials which exists between the plant and surrounding atmosphere. The transpiration from the vegetated surface, summed with the evaporation from the soil is called evapotranspiration. The root water uptake contributes to the soil desaturation and causes an increase of the suction in the soil. Consequently, the apparent cohesion, which is a product of the degree of saturation and suction, increases and causes an enhancement of the soil strength.

The complexity of the processes which take place in the vadose zone and non-uniformity of plants growing on slopes, increase the difficulty of numerical modelling of rainfall-induced landslides of vegetated slopes. Therefore, an appropriate model, which allows combining the most important features resulting from vegetation presence, has to be used.

This paper aims to show the numerical model which is capable of modelling complex, multiphase problems. Moreover, a case study slope is calculated using Comes-Geo [3–6] finite element code, in which the model is implemented.

2 Theory

The coupled hydro-mechanical model for soil reinforced with roots (CHMR), used in numerical modelling of rainfall-induced landslides of vegetated slopes is based on the modified Cam-clay model for unsaturated soils, developed by Tamagnini [11]. In this model, a double-hardening mechanism is incorporated. Therefore, the yield surface can shrink or expand not only due to the increment of the plastic volumetric strains, but also when wetting or drying of the soil occurs. In the extension of the model proposed by Switala [8] and Świtała et al. [9], the expansion of the yield surface occurs also when progressive activation of roots strength takes place. Roots activation proceeds with increasing strains in the soil-root composite, until the critical point is reached and failure occurs (roots start to brake or are pulled-out of the soil mass).

The effective stress formulation used in the model is based on the generalized Bishop's stress, with the weighting factor χ replaced by the degree of saturation [7]. In the following formulation, suction is defined as the difference between air and pore water pressure: $(u_a - u_w)$.

$$\boldsymbol{\sigma}' = \boldsymbol{\sigma} - u_a \mathbf{I} + S_r (u_a - u_w) \mathbf{I} \tag{1}$$

where $\boldsymbol{\sigma}'$ is the effective stress tensor and $\boldsymbol{\sigma}$ is the total stress tensor, and $\mathbf{I}$ is the unit tensor.

The shape of the yield surface f of the CHMR model, which coincides with the plastic potential g (associated flow rule) is given by the following equation:

$$f = g = q^2 - M^2 \left[p' \left(p'_c - p' \right) \right] = 0 \tag{2}$$

where M is the slope of critical state line and p'_c is the preconsolidation pressure.

The multiple-hardening of the soil-root composite is reflected in the formulation of the increment of the preconsolidation pressure. This increment consists of the three components, each of which is responsible for different hardening mechanism [8]:

$$\dot{p}'_c = \dot{p}'_{c(SAT)} + \dot{p}'_{c(UNSAT)} + \dot{p}'_{c(ROOT)} \tag{3}$$

The components of the multiple-hardening function, given by Eq. 3, are:

a. $\dot{p}'_{c(SAT)} = p'_c \dot{\varepsilon}^p_v (1 + e) / (\lambda - \kappa)$, is the size of the yield surface, defined by the volumetric plastic strain increment ($\dot{\varepsilon}^p_v$). This definition is adopted from the original Modified Cam-clay model, developed for the saturated conditions. The parameter λ is the slope of the normal compression line in the compression plane, and κ is the slope of the swelling line in the unloading-reloading plot in the compression plane. Void ratio is denoted with e.
b. $\dot{p}'_{c(UNSAT)} = -bp'_c \dot{S}_r$, is the hardening mechanism governed by the changes in the degree of saturation S_r. The evolution of the yield surface due to these changes may occur even if the current stress lays in the elastic domain [11]. Its rate is controlled by the constitutive parameter b.
c. $\dot{p}'_{c(ROOT)} = R_p p'_c e \dot{\varepsilon}_r$, is the expansion of the yield surface caused by the activation strain increment $\dot{\varepsilon}_r$ (increment of volumetric and deviatoric strains). It defines the level of the activation of the strength of the roots in a root zone. The rate of these changes is governed by a new constitutive parameter R_p. The value of this parameter is dependent on the plant's characteristics and can be obtained from the numerical modelling of triaxial or direct shear tests (fitting of numerical to the experimental results).

The numerical model is implemented in the finite element code Comes-Geo. An implicit integration of the model is done using the return mapping technique by

Borja and Lee [1], with a modified elastic predictor stage. This modification allows change of the dimensions of the yield surface (expansion or shrinkage) in the elastic predictor step. This change may happen due to the drying/wetting of the soil or due to the activation of roots strength or their failure. Therefore, the trial preconsolidation pressure is given by the following formula:

$$p_{\mathrm{c}n+1}^{\mathrm{tr}} = p_{\mathrm{c}n} \exp\left(-b\Delta S_{\mathrm{r}n+1} + R_{\mathrm{p}} e \Delta\varepsilon_{\mathrm{r}n+1}\right) \tag{4}$$

Further details regarding model derivation and implementation can be found in the work by Switala [8] and related publications [9, 10].

3 Vegetated Slope - Case Study

3.1 Introduction

In this section, a case study from Matzen in Lower Austria is presented. This research is conducted in cooperation with the J. Krismer GmbH company from Austria. The most unfavourable cross-section of the slope in Matzen is chosen to be modelled. A rainfall-induced landslide is simulated using Comes-Geo code. Two cases are considered: vegetated and non-vegetated. Rainfall of a constant intensity is applied to the top surface of the both slopes and time after which landslide occurs is monitored.

3.2 Non-vegetated Slope

The geometry of the modelled slope is based on the data from the field documentation. The slope surface has a complex structure. In the steepest part, the inclination reaches almost 80°. Therefore, it is evident that high negative pore pressure must exist in the soil, to assure its stability in such difficult conditions, prior the application of stabilisation measures.

The geometry of the slope and the finite element discretization is presented in Fig. 1. The slope is discretized with isoparametric, quadrilateral, 8-nodes finite elements. Material parameters of the model are listed in Table 1. Due to the lack of experimental data, values of some parameters are assumed, taking typical values for loess. Soil water retention curve is defined by the following formula:

$$S_{\mathrm{r}} = 1 - \frac{1}{\beta \log\left(u_{\mathrm{c}}/a\right)} S_{\mathrm{sat}} \tag{5}$$

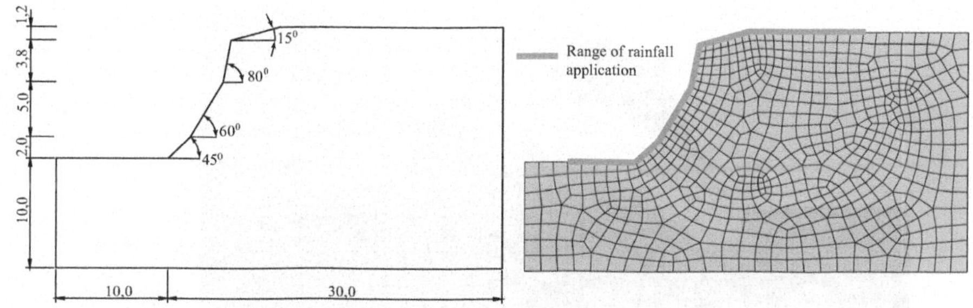

Fig. 1 Geometry of the slope in Matzen (left-hand side) and its finite element discretization (right-hand side) [8]

Table 1 Soil parameters used in the numerical model of the case study in Matzen, performed in *Comes-Geo*

Soil density	ρ_s	2670.0	kg/m^3
Initial young modulus	E_{mod}	20000.0	kPa
Poisson's ratio	ν	0.3	–
Initial void ratio	e_0	1.3	–
Intrinsic permeability of the soil	K	1.0e–13	m^2
Initial preconsolidation pressure	p_{c0}	100.0	kPa
Plastic compressibility	λ	0.111	–
Elastic compressibility	κ	0.040	–
Parameter for partial saturation	b	10.0	–
Initial degree of saturation	S_r	0.55	–
Slope of the CSL	M	1.0	–
Water density	ρ_w	1000.0	kg/m^3

where a is the air entry value, β is the fitting parameter and S_{sat} is the maximum degree of saturation of the soil. This relation is adopted from the work of Francois [2]. The values of the parameters are: $\beta = 5.5$, $a = 10{,}000$ Pa, and $S_{sat} = 0.95$.

The calculations done in *Comes-Geo* result in the wetting collapse of the soil. The numerical calculations are performed in three steps. First the geostatic stresses of the system are initialized to reach the equilibrium (in this step gas pressure $u_g = 101.3$ kPa, capillary pressure $u_c = 0$, $T = 293.15$ K, bottom of the model is constrained in the vertical direction y and sides are constrained in the horizontal direction x). After the geostatic step, the capillary pressure of $u_c = 110$ kPa is assigned to all the nodes in order to reach a uniform degree of saturation of $S_r = 0.55$ of the soil mass. In the next step, the slope is subjected to rainfall infiltration of the constant intensity of 18 mm/h. However, this value is applied only to the flat surfaces of the model, whereas rainfall intensity on inclined surfaces is a fraction of the maximum value, equal to 18 mm/h multiplied by the cosine of the slope angle. The range of the rainfall application

Fig. 2 Contour of the displacements norm obtained after approximately 40 h of rainfall. Displacements are given in [m]

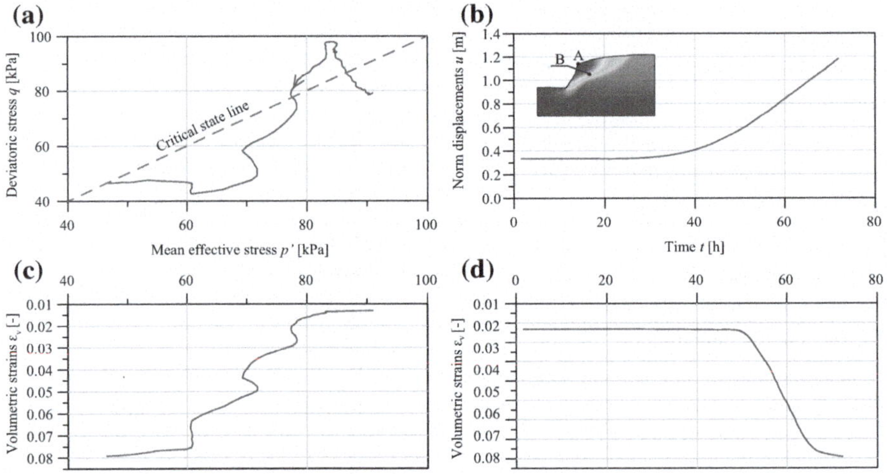

Fig. 3 Results from the numerical analysis of the bare slope in Matzen: Numerical results: **a** stress path in a point (B) in the middle of the sliding mass, **b** total displacement (Euclidean norm) of the point (A) at the top of the slope, **c** volumetric strains versus mean stress in the point B, **d** volumetric strains of the point B. Location of points A and B is marked in the inlet of (**b**)

is shown in the left hand side of the Fig. 1. The failure occurs approximately after 40 h when the top layer of the soil is fully saturated and significant displacements develop. Contour of the displacements norm is shown in Fig. 2.

The cumulation of displacements, volumetric strains as well as the stress path of the soil are presented in Fig. 3.

The stress path reflects changes in the mean effective stress together with the changes in deviatoric stress. When the stress path approaches the CSL, the failure occurs and the soil collapses ($p' \simeq 78$ kPa). In Fig. 3c, the accumulation of the volumetric strains can be observed, accompanied by the decrease in the mean effective

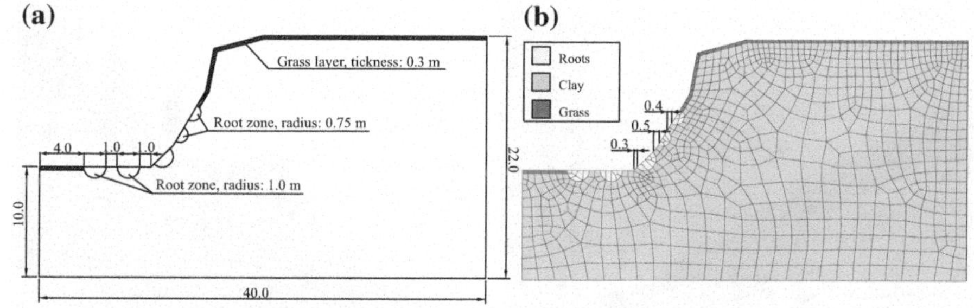

Fig. 4 **a** geometry of the slope, **b** finite element discretization of the numerical model of the vegetated slope in Matzen. All the dimensions are in [m]

stress. The analysis of the Fig. 3b, d shows that the rapid increase in the norm of the displacements after approximately 40 h of rainfall corresponds well to the accumulation of the volumetric strains in the shear zone. This is a visible indicator of the slope failure.

3.3 Vegetated Slope

In order to improve the stability of the slope, and to protect the surface against erosion, soil-bioengineering methods can be used. An example of vegetation placement on the slope is proposed in this section. The dimensions of the model, and finite element discretization are presented in Fig. 4.

The situation on the slope before the reconstruction and stabilisation is shown in Fig. 5. It can be noticed that although the upper part of the slope is very steep, it is stable. The reason for that is the presence of plant roots in the slope which bind the soil together, reinforce it, and possibly develop an enhanced unsaturated regime. All these factors contribute to the significant increase of the slope stability.

Different types of trees and shrubs can be planted at the toe of the slope, where their contribution to the stability is the most effective. Smaller and lighter plants can stabilise higher and steeper parts. Finally, on the crest, only a grass layer covers the surface. An appropriate plant distribution on the slope is vital to assure the effective stabilisation. In high and steep parts of the slope, large plants may reduce the stability due to the significant additional surcharge and increased exposure of the higher vegetation to the action of wind forces. Contrarily, small light plants like grass or herbs, which cover upper steep parts of the slope, protect the soil surface from erosion.

The shapes of the root zones in this example are assumed to be semi-circular. The impact of vegetation on slope stability is investigated, taking into account both the mechanical and hydrological contribution. The evapotranspiration can be modelled

Fig. 5 The slope in matzen before the reconstruction

in Comes-Geo as a negative flux of water vapour, applied to the surface nodes within the root zones.

Steps of the numerical analysis are similar to those from the calculations of the bare slope case. However, before the rainfall step (see steps sequence in Sect. 3.2), the evapotranspiration is applied as an additional step. The value of evapotranspiration from trees and shrubs, with semi-spherical root architectures, is assumed to be equal to 3.6, and 0.22 mm/h from the grass surface. The root constitutive parameter R_p is set to 10.0 within semi-spherical root zones and 4.0 within grass root layer. At the end of the drying step, significant matric suction develops in the vicinity of the root zones. This contributes to the increase of the apparent cohesion.

After the evapotranspiration step, rainfall of an intensity of 18 mm/h (or smaller for inclined parts of the surface) is applied to the nodes on the top surface of the model. In this case, the time after which a landslide occurs is significantly prolonged when compared to the non-vegetated case. The comparison of the development of displacements in time for the case of the bare and vegetated slope is shown in Fig. 6. In this figure, the extreme accumulation of displacements of the bare slope is an effect of about 40 h of rainfall. This is an indicator of the failure occurrence. When vegetation is planted on the slope, the time when slope is stable is prolonged. Around 100 hours after the start of the rainfall, the initiation can be noticed, but the cumulation of displacements is still not significant. The contour of displacements at the end of calculations is shown in Fig. 7.

4 Conclusions

The presented paper highlights the importance of the presence of vegetation on slopes. The developed model allows simulation of rainfall-induced landslides and comparison between the stability of vegetated and non-vegetated slopes. What is important, the model takes into account and couples the most important factors which influence the slope stability, namely mechanical root reinforcement and evapotranspiration.

It is clear that bio-engineered slopes can sustain longer rainfall periods and, if a landslide occurs, the cumulation of displacement is not that rapid as in the case of a bare slope. Safety enhancement is defined in reference to the stability of a non-vegetated slope, as the delay of the occurrence of a rainfall-induced landslide. In this way, different plant placement strategies on the slope can be tested, resulting in the most efficient reinforcement. Moreover, it is also noteworthy how serious can be the consequences of the vegetation removal, especially from steep forested slopes.

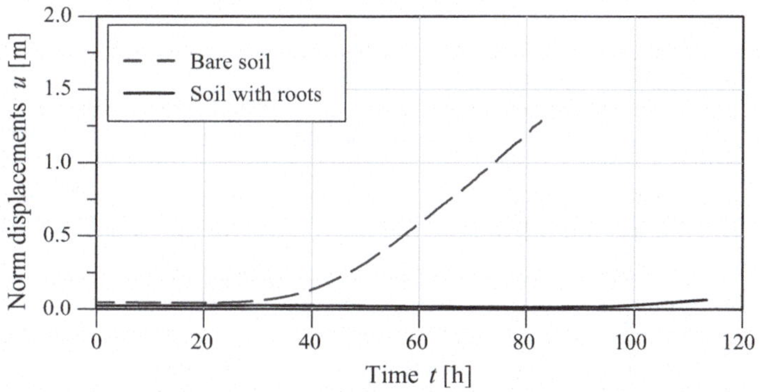

Fig. 6 Comparison of displacements measured on the top of the slope for a bare slope and for the vegetated slope, considering mechanical and hydrological effects of vegetation which contribute to slope stability

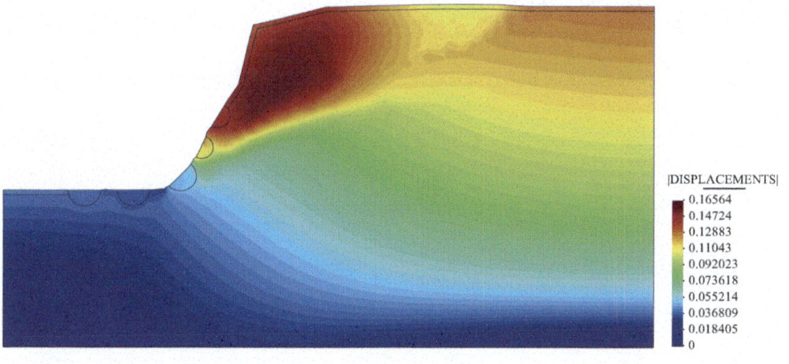

Fig. 7 Contour of displacements of the vegetated slope in Matzen at the end of calculations. Values of displacements are given in [m]

Acknowledgements This research was undertaken as a part of the MUMOLADE (Multiscale Modelling of Landslides and Debris Flows) Initial Training Network, funded by the Marie Curie Actions and European Union's Seventh Programme Research Framework, under REA grant agreement n° 289911. Moreover, we would like to thank the University of Padova for providing the Comes-Geo code and assistance in its use, and J. Krismer GmbH for providing the case study field documentation. We thank also to the European Commission for the financial support to the project Geohazards: Risk Assessment, Mitigation and Prevention (GEORAMP) (Project ID: 645665) under Horizon 2020.

References

1. Borja, R.I., Lee, S.R.: Cam-clay plasticity, part I: Implicit intergration of elasto-plastic constitutive relations. Comput. Methods Appl. Mech. Eng. **78**, 49–72 (1990)
2. Francois, B.: Thermo-plasticity of fine grained soils at various saturation states: application to nuclear waste disposal. Ph.D. Thesis, Ecole Politechnique Federale de Lausanne, Switzerland (2008)
3. Gawin, D., Schrefler, B.A.: Thermo-hydro-mechanical analysis of partially saturated porous materials. Eng. Comput. **26**(3), 245–266 (2009)
4. Sanavia, L.: Numerical modelling of a slope stability test by means of porous media mechanics. Eng. Comput. **37**, 331–348 (2009)
5. Sanavia, L., Francois, B., Bartolotto, R., Luison, L., Laloui, L.: Finite element modelling of thermo-elasto-plastic water saturated porous materials. J. Theor. Appl. Mech. **38**(1–2), 7–24 (2008)
6. Sanavia, L., Pesavento, F., Schrefler, B.A.: Finite element analysis of non-isothermal multiphase geomaterials with application to strain localization simulation. Comput. Mech. **37**, 331–348 (2006)
7. Schrefler, B.A.: The finite element method in soil consolidation (with application to surface subsidence). Ph.D. Thesis, Univeristy College of Swansea, UK (1984)
8. Switala, B.M.: Analysis of slope stabilisation by soil bioengineering methods. Ph.D. Thesis, University of Natural Resources and Life Sciences, Vienna, Austria (2016)
9. Świtała, B.M., Askarinejad, A., Wu, W., Springman, S.M.: Experimental validation of a coupled hydro-mechanical model for vegetated soil. Géotechnique **68**(5), 375–385 (2018)
10. Świtała, B.M., Wu, W.: Numerical modelling of rainfall-induced instability of vegetated slopes. Géotechnique. http://dx.doi.org/10.1680/jgeot.16.P.176
11. Tamagnini, R.: An extended cam-clay model for unsaturated soils with hydraulic hysteresis. Geotechnique **3**, 223–228 (2004)

Behaviour of Reinforced Sand with Synthetic Fibres in a Centrifuge

Rick Veenhof and Wei Wu

Abstract This study describes the effects of adding randomly distributed polyolefine monofilamente fibres on the behaviour of medium-fine sand in a loose condition. Direct shear tests and centrifuge tests were performed on sand specimens with a void ratio (e) of 0.9 and water content of 11.5%. The reinforced sand had a fibre content of 0.25%. An increase in the peak shear strength with a factor of 1.085 on average is observed. Centrifuge tests show that for slopes that are reinforced the period until failure is extended. The location of shear band formation and patch displacement behaviour indicate that the design of slope reinforcement has a significant effect on the failure behaviour.

1 Introduction

The use of fibres as a way to reinforce the earth has been applied widely and is generally considered to be effective [5]. Theoretical models, laboratory tests and numerical simulations have aided in gaining insight in the process underlying the gain in soil strength through this reinforcement. A lot of information has been gained through the application of direct shear tests to estimate soil strength parameters and behaviour [2–4, 6, 7, 10]. A general conclusion is that the added value of fibres lie in its high tensile strength and extendibility properties, which result in a more durable solution in which the loss of post peak strength is smaller compared to more inextensible inclusions such as metal strips or cement [9, 11]. Less research

R. Veenhof (✉) · W. Wu
Institute of Geotechnical Engineering, University of Natural Resources and Life Sciences, Feistmantelstrasse 4, 1180 Vienna, Austria
e-mail: rick.veenhof@boku.ac.at

W. Wu
e-mail: wei.wu@boku.ac.at

W. Wu (ed.), *Recent Advances in Geotechnical Research*,
Springer Series in Geomechanics and Geoengineering,
https://doi.org/10.1007/978-3-319-89671-7_17

is done on how reinforced soil behaves under slope conditions despite the fact that many areas are prone to slope failure. In this research we are therefore interested in understanding the behaviour of reinforced soil with fibres on slopes, to contribute to an improved design of man-made and natural slopes. Below, a review is presented of several laboratory tests involving direct shear and geotechnical centrifuge tests. The soil and fibres characteristics for the direct shear tests are given in Tables 1 and 2. And for the centrifuge tests in Tables 3 and 4.

Gray and Ohashi [6], were one of the first, who did experimental tests with different types of fibres in sand to obtain a reinforcement effect. Direct shear test results showed that fibre reinforcement increased the peak shear strength and limited post peak reductions in shear resistance. In particular they found that shear strength increases were directly proportional to the concentration or area ratio of fibres, the greatest for initial fibre orientation of 60° and similar in loose and dense state. The findings were in line with theoretical predictions based on a force equilibrium model.

Ibraim and Fourmont [7], examined the effect of polypropylene fibres on the mechanical behaviour of reinforced sand by conducting direct shear tests. They varied fibre densities using different combinations of void ratios and effective normal stresses. Results show that with increasing fibre content the increase of peak shear strength was almost linear at low normal effective stress and a high void ratio, and approached a limiting value at higher normal stress and a low void ratio.

Eab et al. [3], studied the shear strength of a soil sample with 2% of fibres by making use of the direct shear apparatus and this showed an increase of cohesion by 14 KPa and friction angle increase of 3°. They compared these results with Vangbunkong et al. [13], who states that the increase in shear strength mainly arises from the cohesion parameter.

Noorzad and Zarinkolaei [10], Eldesouky et al. [4] and Claria and Vettorelo [2], published all recently about fibre reinforcement sand tested via direct shear tests. [10], did 40 direct shear tests to analyse the effect of the parameters; fibre content, length of fibre and normal and confining pressure. They concluded that the addition of fibres improves shear strength parameters (c, ϕ). It increases the peak shear strength and axial strain at failure, and limited the amount of post/peak reduction in shear resistance. Noorzad and Zarinkolaei [4], conducted 108 specimens of fibre-reinforced sand using direct shear tests to investigate the effects of the fibres content, relative density, normal stress and moisture content on shear strength and volumetric change behaviour. The fibres inclusion improves the peak and post-peak shear strengths of dry sand by up to 50 and 30%, respectively. On the other hand, the fibres inclusion increases the sand dilation. Adding moisture to the sand-fibre specimen reduces the fibres effect on the peak and post-peak shear strengths. The peak shear strengths of the moist specimens are affected by the relative density increase more than they are affected by the fibres inclusion. The post-peak shear strength drop is reduced by the introduction of moisture to fibre-reinforced sand. Changing this moisture content on the dry side does not have any noticeable impact on the peak shear strength, the post-peak shear strength and dilation of unreinforced and fibre-reinforced sand. Lastly they concluded that, dry loose 0.5% reinforced sand achieves the same peak shear strength of moist very dense unreinforced sand, yet at more than double the horizontal

Table 1 Soil characteristics of [2–4, 6, 7, 10] used during direct shear tests

Article	Type of sand	D_{50} (mm)	G_s (kN/m^3)	C_u (–)	C_g (–)	e_{min} (–)	e_{max} (–)
Claria and Vettorelo	Medium to fine siliceous sand	0.50	2.68	2.4	1.4	–	–
Eab et al.	Edosaki sand	–	2.65	–	–	0.87	1.29
Eldesouky et al.	Poorly graded siliceous sand	–	2.64	1.5	–	0.48	0.72
Gray and Ohashi	Clean quartz beach sand	0.23	–	1.5	–	0.50	0.73
Ibraim and Fourmont	Hostun RF (S28)	0.32	2.65	1.7	1.1	0.62	1.00
Noorzad and Zarinkolaei	Poorly graded sand (SP)	0.19	2.74	1.7	–	0.58	0.86

Table 2 Fibres characteristics of [2–4, 6, 7, 10] used during direct shear tests

Article	Content (%)	τ (MPa)	⌀ (mm)	ϕ (°)	ϕ with (°)	c (kPa)	c with (kPa)
Claria and Vettorelo	1.0–4.0	–	–	36.0	44.0	–	–
Eab et al.	2.0	–	0.039	28.6	31.5	4.8	18.9
Eldesouky et al.	0.50–1.0	350	0.05	40.0	45.0	–	–
Gray and Ohashi	0.25– 0.50	–	–	31.0	39.0	–	–
Ibraim and Fourmont	0.30	300	0.10	31.4	33.6	4.5	8.9
Noorzad and Zarinkolaei	0.25–1.0	400	0.023	40.0	43.0	0.0	4.0

displacement. One of the latest researches about reinforcement of sand is by Claria and Vettorelo [2], they describe the mechanical behaviour of loose sand reinforced with meshed and smooth synthetic fibres. They concluded that the addition of fibres increases the shear strength at large strain level, the ductility of the reinforced sand and the internal frictional angle. The shear strength of the reinforced sand seems to increase indefinitely with fibre length. However, if failure is defined in terms of the maximum tolerable strain, fibres longer than 10 mm do not produce significant improvement in shear strength. And finally, they conclude that fibre content and fibre length have a significant influence on the stress-strain and strength behaviour of the reinforced sand. The addition of fibres produces an increment in the shear strength and ductility but a loss of stiffness.

Little research has been done related to centrifuge tests with reinforcement of sand. Eab et al. [3] and Wang and Brennan [14], are one of the first to apply fibres-reinforced sand in a geotechnical centrifuge. Wang and Brennan [14], did two dynamic centrifuge tests, one test acting as a reference with clean sand and the other entirely reinforced with fibres-sand mixture, both subjected to an earthquake motion. The shear modulus and sand matrix is increased with the existence of fibres especially in lower overburden stress. Eab et al. [3], describes a soil mixture with polyester fibre

Table 3 Soil characteristics of [3, 14] used during geotechnical centrifuge tests

Article	Type	D_{50} (mm)	G_s (kN/m^3)	C_u (–)	C_g (–)	e_{min} (–)	e_{max} (–)
Eab et al.	Edosaki sand	–	2.65	–	–	0.87	1.29
Wang and Brennan	HST-95 sand	0.13	2.63	2.25	–	0.46	0.80

Table 4 Fibres characteristics of [3, 14] used during geotechnical centrifuge tests

Article	Type	Content (%)	τ (MPa)	E (MPa)	SG (MPa)	⊘ (MPa)
Eab et al.	Polypropylene	2.0	–	–	–	0.039
Wang and Brennan	Polypropylene	–	200	–	0.91	0.10

slope failure induced by rainfall in a geotechnical centrifuge. They concluded that a slope whose surface was reinforced, exhibited small deformation without failure under heavy rainfall (intensity 20 mm/h) for a duration of 15 h at prototype scale.

The available literature shows that much research has been done with fibre-sand mixtures in direct shear tests and less in geotechnical centrifuge tests. Therefore to gain more insight in the behaviour of sand in loose conditions centrifuge tests have been made to compare sand reinforcement under slope failure. To put the sand reinforcement mixtures in perspective, here is chosen to compare our direct shear results with existing literature to see how this reflects the general outcome of adding fibres to sand. The aim of our study is to provide insight in the effect of fibres on soil strength under a given slope geometry, with specific density and fibre content.

Presented in this paper are the results of direct shear tests and centrifuge tests. Specifically, the objective of this paper was to examine the effect of randomly distributed fibres on slope stability using/experimenting with different reinforcement designs. In order to place the results in perspective and build on a wider body of previous research, direct shear tests are done in addition to the centrifuge tests.

2 Methods

2.1 Direct Shear Box Tests

The direct shear apparatus at the Institute of Geotechnical Engineering, University of Natural Resources and Life Sciences, Vienna, BOKU is employed to determine the shear strength of the tested specimens. The shear box dimensions are 160 $\times$130 $\times$ 55 mm and is divided horizontally into two halves. The sand that was used is classified as well-graded, medium-fine sand with minor fraction of silt according to the ÖNORM EN ISO 14688-2 (2013 11 15). The soil characteristics are given

Table 5 Soil Characteristics

D_{10} (mm)	D_{50} (mm)	C_u (–)	C_c (–)	e_{min} (–)	e_{max} (–)	ρ_s (g/cm^3)
0.08	0.252	3.7	1.2	0.56	1.05	2.70

(a) Original size of fibres (b) Zoom in of fibres to 1 mm

Fig. 1 Polyolefine monofilament fibres

in Table 5. Around 300grams was used to fill the direct shear box with unreinforced sand, with a water content of 11.5%. The box with soil sample is placed in an automated direct shear apparatus and sheared at a displacement rate of 0.3 mm/min. During testing, the lower section of the box was moved forward at a constant velocity while the upper section of the box remained static. Four different normal stresses (50, 100, 200 and 400 kPa) were employed.

For the reinforced sample, polyolefin fibres were used, (Avenarius Agro Fibrofor multi 127) with a diameter of 32 μm, a length of 12 mm (±5%) and a tensile strength of >270 N/mm^2, specific gravity of 0.91 [1] (Fig. 1). The content of fibre utilized in the test was 0.25% of the weight of the dry sand.

2.2 *Centrifuge Modelling*

The experiments were done using the beam-type centrifuge at the Institute of Geotechnical Engineering, University of Natural Resources and Life Sciences, Vienna, BOKU. The centrifuge is characterized by a maximum radius of 1.30 m from the base to swinging platform and a loading capacity of 10 g-t, with a maximum g-level of 200g [12]. Results from earlier centrifugal experiments performed by Idinger [8] showed that for unreinforced sand with density ranges of 1.40–1.53 g/cm^3 a minimum slope angle of 55° was necessary in order to observe failure initiated by

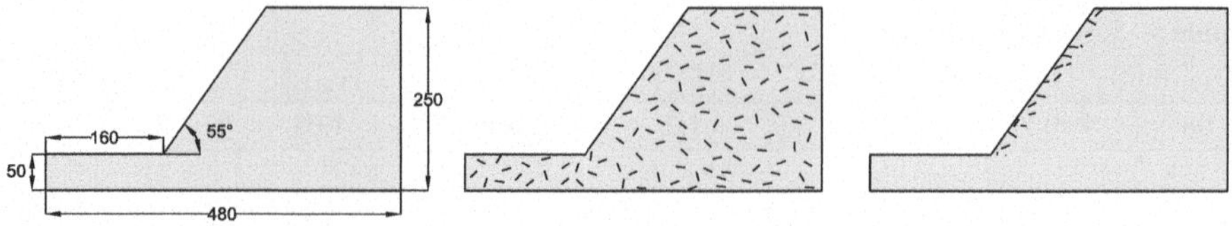

Fig. 2 Geometry of the slopes [mm] and overview of randomly distributed fibres, whole slope reinforced and 20 mm top slope layer

self-weight. The same sand as used during the direct shear tests was used well-graded, medium-fine sand with minor fraction of silt. Given this information the following set-up was used in this study. One reference test without reinforcement, one test using completely reinforced soil and one test with a 20 mm reinforced layer at the slope were performed under an angle of 55° (Fig. 2). While it is hypothesized that from a strengthening perspective complete reinforcement will perform best, it is also acknowledged that from a practical and/or cost perspective complete reinforcement is not always an option. Hence, both tests are performed in this study. The relative density of the sand is 1.40 g/cm^3 with a water content of 11.5%. Fibres were added via mixing with a mixing drill. Similar to the direct shear test, reinforced sand with 0.25% fibre content was used. The model slope was constructed inside a steel box with inner dimensions of 480 $\times$ 155 $\times$ 450 mm. In order to obtain the desired density the slope is built up layer for layer via the moist tamping method using a compaction plate and hammer. In case of the reference soil a slope is created using wooden templates on either side of the soil profile and by cutting away the redundant soil, Fig. 2. In the case of the reinforced slope the whole soil-sample consisted of the randomly distributed sand-fibre composite. Related to the 20 mm reinforced slope soil, layering was divided into reinforced and non-reinforced soil. To prevent shearing stress at the side wall of the container box, silicon oil was sprayed on before compaction.

The main component to trigger the failure of the reinforced slope is gravity, to achieve this an continuous increase of the gravity-induced stress field, is applied. The stress increase in the model slope corresponds to the increase of the prototype slope height. These so called g-up tests are designed to end when the reinforced slope failed or 70g is reached. The increase of centrifugal acceleration is operated at a rate of 0.1 rpm per second by a computer program. The movement of the slope was captured by an digital still camera ($Canon\ PowerShot\ G10$) inside the centrifuge, close to the model box. Afterwards soil deformation was studied with PIV analysis [15].

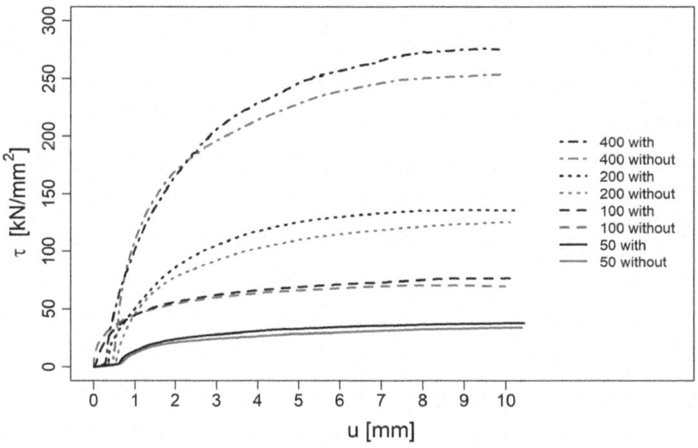

Fig. 3 Results direct shear test

3 Results

3.1 Direct Shear Box Tests

Presented here is the outcome of the direct shear box tests for unreinforced and reinforced soil. Figure 3 shows the obtained shear strain-curve, with increasing normal stress, the black lines are with reinforcement, the gray lines without. The higher the normal stress applied, the higher the observed peak shear strength of the materials. Under all conditions an increase in peak shear strength for the reinforced material is observed compared to unreinforced material. In absolute terms, the gain in peak shear strength becomes larger under increasing normal stresses. Dividing peak shear strength of reinforced soil by peak shear strength of unreinforced soil gives ratio's of 1.082, 1.087, 1.084 and 1.087, for 50, 100, 200 and 400 kPa, respectively. Figure 4 shows the failure envelopes for unreinforced (gray) and reinforced (black) soil samples. The envelopes are linear and the failure envelope moves upward when adding fibres. The internal friction angle (ϕ) equals 31.4° for unreinforced and 33.3° for reinforced. The cohesion is 4.6 [kN/m^2] for unreinforced and 6.7 [kN/m^2] for reinforced soil.

3.2 Centrifuge Tests

In this section the results of the centrifuge tests are presented. The photos show slope conditions for unreinforced (a), 20 mm reinforced (b) and complete reinforced sand (c) (Fig. 5). Failure occurred at the unreinforced slope after 24g, for the 20 mm

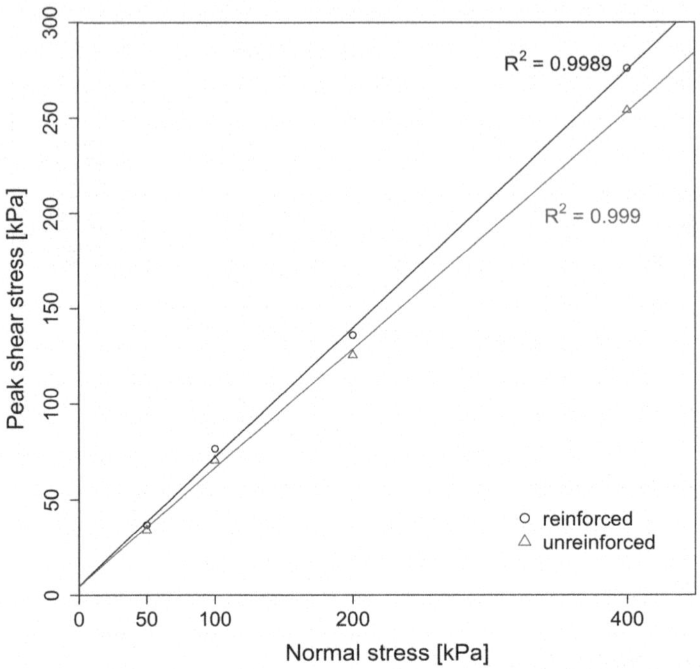

Fig. 4 Failure envelopes

reinforced slope after 29g. The slope that was completely reinforced with fibres did not fail when subjected up to 70g. The total maximum shear strain profiles for the three centrifuge tests are shown in Fig. 6 at 20g. This level is in all cases pre-failure. Most similarities are observed between the unreinforced and 20 mm reinforced slope. Maximum shear strain values are observed up until a value of 8 and in both cases a clearly observable shear band is apparent. The difference between the 20 mm reinforced and the unreinforced slope is that the shear band is formed at a larger, angle almost parallel to the slope itself. On the contrary, the completely reinforced slope does not show a clear shear band and experiences a more uniform maximum shear strain up to a value of 3.

Figure 7 illustrates the horizontal displacement of the crest of the slope for all three tests under increasing g-level. Exponential displacement is observed for both the unreinforced and 20 mm reinforced patch. The process of displacement of the 20 mm reinforced slope until failure is however extended by almost a factor of two. The completely reinforced slope shows almost linear behaviour for the current g-levels.

4 Discussion

The outcome that peak shear strength of reinforced soil increases compared to unreinforced soil has been observed before Ibraim and Fourmont [7]. To see how the results of this study are compared to previous research, the internal friction angle, cohesion and peak shear strength ratio are analyzed. In case of multiple tests, the test set-up that matched the set-up of this research closest was selected. Ibraim and Fourmont [7], found for a unreinforced soil sample (void ratio of 0.9, moisture content 10%) a friction angle of 31.4° and a cohesion of 4.5 kPa. The values found in this study were respectively 31.4° and 4.6 kPa. For reinforced soil (fibre content of 0.3%) values of 33.6° and 8.9 kPa were obtained, versus 33.3° and 6.7 kPa. Based on the results of Noorzad and Zarinkolaei [10], who observed that the angle of internal friction and the cohesion gradually increase with increasing fibre content, the

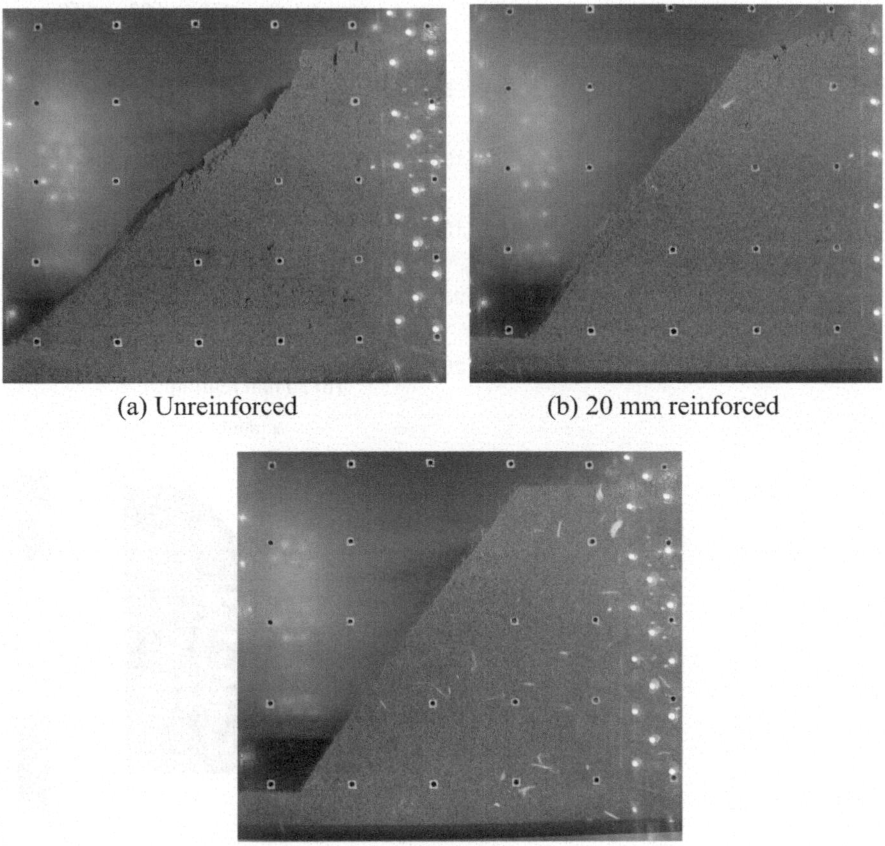

(a) Unreinforced (b) 20 mm reinforced

(c) Reinforced

Fig. 5 Photos of the slope conditions

Fig. 6 Total maximum shear strain at 20g

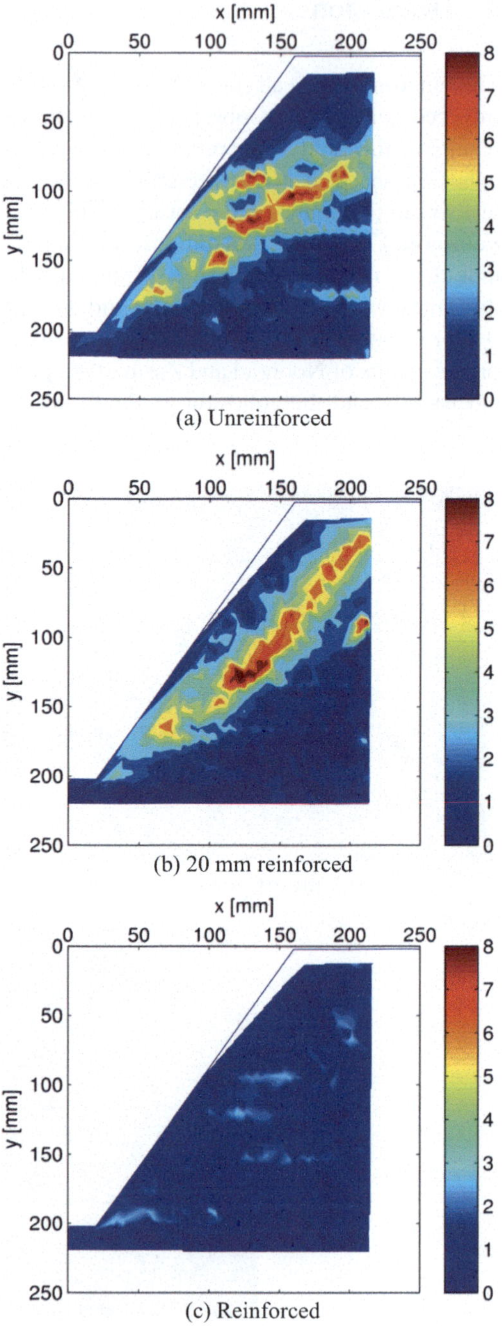

(a) Unreinforced

(b) 20 mm reinforced

(c) Reinforced

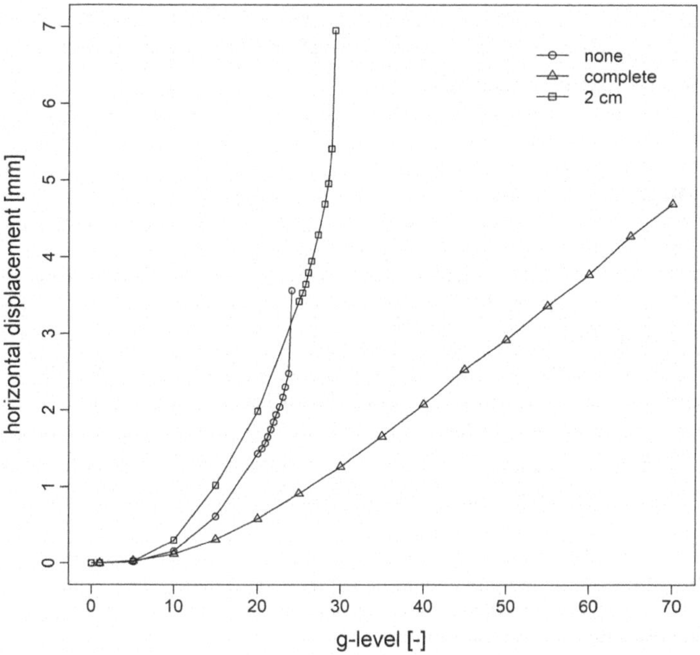

Fig. 7 Single patch displacement

findings of this study seem to fit the general pattern. The conclusion that peak shear strength ratio increases as fibre content increases as Noorzad and Zarinkolaei [10] suggest as well, is not supported by the results found in this study. The observed ratios of 1.082, 1.087, 1.084 and 1.087 (1.09 on average) do not show a clear trend. Peak shear strength ratio's that were extracted from Fig. 8 in [7] equaled 1.15 on average, and 1.1, 1.13, 1.15, 1.2 at applied normal stresses of 55.3, 106.4, 208.5 and 310 kPa, respectively. Although measurement errors cannot be excluded, these observed show an increase in trend. Either an increase or decrease in ratio would stand in contrast with the assumption of linearity when constructing the failure envelope (see Fig. 4). The average peak shear strength ratio reported by [10] for a 0.25% fibre content, void ratio (unknown) equals 1.19. Test results within the same test-setup indicate that peak shear strength continues to increase when fibre content % increases. The current comparison between test does not yet confirm this, indicating that the outcome is quite sentative to the preparation of the sample.

The results of the centrifuge test for unreinforced material are in line with the results of Idinger [8], who found a g-level of 22 for similar soil parameters (compared to 24g in this test). Since the authors are unaware of any studies with the current set-up that have been done using the centrifuge, a direct comparison is not possible. Current observations indicate that the timing of the failure and the location of the shear band are dependent on the design of reinforcement. While reinforcement postpones the onset of the failure, the weakest location within the slope seems to be the transition of the reinforced material to the unreinforced slope. Further research could therefore focus on improving this transition.

5 Conclusion

Both direct shear tests and centrifuge tests were conducted to investigate the effect of fibres on soil strength and slope failure. For loose sand with a void ratio of 0.9 and a water content of 11.5%, reinforced with 0.25% fibres an increase in the peak shear strength with a factor of 1.085 on average is observed. Centrifuge tests show that for slopes that are reinforced the period until failure is extended. The design of slope reinforcement has a significant effect on the failure behaviour as was illustrated by the location of shear band formation and patch displacement behaviour. Further research is needed to determine the optimal design of reinforced slopes.

Acknowledgements The research carried out in this paper is partly funded by the project 'GEORAMP' within the RISE programme of Horizon 2020 under grant number 645665. The first author would like to acknowledge the financial support from the Otto Pregl Foundation for Fundamental Geotechnical Research in Vienna.

References

1. AvenariusAgro. Fibrofor - faserzusatz multi 127. Technische Information Nr. 1005 / 4 (2015)
2. Claria, J.J., Vettorelo, P.V.: Mechanical behavior of loose sand reinforced with synthetic fibers. Soil Mech. Found. Eng. **53**(1), 12–18 (2016)
3. Eab, K.H., Takahashi, A., Likitlersuang, S.: Centrifuge modelling of root-reinforced soil slope subjected to rainfall infiltration. Géotechn. Lett. **4**(3), 211–216 (2014)
4. Eldesouky, Morsy, and Mansour] Eldesouky, H.M., Morsy, M.M., Mansour, M.F.: Fiber-reinforced sand strength and dilation characteristics. Ain Shams Eng. J. **7**(2), 517–526 (2016)
5. Gao, Z., Zhao, J.: Evaluation on failure of fiber-reinforced sand. J. Geotech. Geoenviron. Eng. **139**(1), 95–106 (2012)
6. Gray, D.H., Ohashi, H.: Mechanics of fiber reinforcement in sand. J. Geotechn. Eng. **109**(3), 335–353 (1983)
7. Ibraim, E., Fourmont, S.: Behaviour of sand reinforced with fibres, pp. 807–818 (2007)
8. Idinger, G.: Experimental study on failure initiation in partially saturated slopes. Ph.D.-Thesis. University of Natural Resources and Life Sciences, Vienna (2016)
9. McGown, A., Andrawes, K., Al-Hasani, M.: Effect of inclusion properties on the behaviour of sand. Geotechnique **28**(3), 327–346 (1978)
10. Noorzad, R., Zarinkolaei, S.T.G.: Comparison of mechanical properties of fiber-reinforced sand under triaxial compression and direct shear. Open Geosci. **7**(1) (2015)
11. Park, S.-S.: Unconfined compressive strength and ductility of fiber-reinforced cemented sand. Constr. Build. Mater. **25**(2), 1134–1138 (2011)
12. Trio-Tech. Model 1231 centrifuge: Operations manual. 94. Manual (1989)
13. Vangbunkong, A., Likitlersuang, S., Takahashi, A.: Investigation of root induced shear strength of vetiver grass reinforced soil. In: 26th KKHTCNN (2013)

14. Wang, K., Brennan, A.: Centrifuge modelling of saturated fibre-reinforced sand. In: ICPMG2014 - Physical Modelling in Geotechnics, pp. 1095–1100 (2014). Edited by Christophe Gaudin and David White
15. White, D., Take, W., Bolton, M.: Soil deformation measurement using particle image velocimetry (PIV) and photogrammetry. Geotechnique **53**(7), 619–632 (2003)

Extension of a Basic Hypoplastic Model for Cohesive Soils

Shun Wang and Wei Wu

Abstract A study on the extension of a basic hypoplastic constitutive model for cohesive soils is presented. Firstly, a translated tensor function is added to the constitutive equation to simulate the effect of cohesion. In addition, the magnitude of the translated tensor is linked to cohesion and internal friction angle of cohesive soils. Secondly, the concept of strain history is incorporated to capture strain softening behaviour of cohesive soils. Finally, the extended model is implemented into a finite element code, and some boundary value problems are simulated to examine its performance.

1 Introduction

Hypoplastic constitutive models are originally proposed for granular materials, which are usually cohesionless soil. In practice, however, most soils are cohesive soils but the attempt to specifically develop hypoplastic models for cohesive soils is still a challenge [1]. Fortunately, the inherent shortcomings can be remedied by extension of the hypoplastic constitutive equations.

Cohesion is the most important behaviour of cohesive soil. There are many ways to simulate the effect of cohesion in hypoplasticity. For instance, Bauer [2] extended a hypoplastic equation by adding an intrinsic pressure to the actual stress to account for cohesion and the effect of strain history. In a similar way, Weifner and Kolymbas [3] introduced a stress-like internal parameter to describe the behaviours of cohesive soils

S. Wang (✉) · W. Wu
Institute of Geotechnical Engineering, University of Natural Resources and Life Sciences, Vienna, Feistmantelstraße 4, 1180 Vienna, Austria
e-mail: shun.wang@boku.ac.at

W. Wu
e-mail: wei.wu@boku.ac.at

W. Wu (ed.), *Recent Advances in Geotechnical Research*,
Springer Series in Geomechanics and Geoengineering,
https://doi.org/10.1007/978-3-319-89671-7_18

for various states of consolidation. Moreover, Wang and Wu [4] updated a hypoplastic model by adding a translated tensor which related to cohesion of clay-like soil to the Cauchy stress tensor. However, the magnitude of the translated tensor is not correctly given. Another approach relates to combining the critical state soil mechanics with hypoplasticity. Following this principle, Mašín [5] proposed a hypoplastic model for clay on the basis of a constitutive equation suggested by Niemunis [6], and Huang [7] suggested a hypoplastic model for clay based on the constitutive model proposed for sands by Gudehus [8] and Bauer [9]. However, the aforementioned approach is usually sophisticated.

On the other hand, strain softening is another important feature for cohesive soils in the post-failure condition. The stress-strain response of most geotechnical materials such as concrete, rocks, and some soils are observed to undergo a softening with strain history. However, to the knowledge of the authors, there has been no reported study that incorporated strain history in hypoplastic models. A possible way is to add a scale function that includes the strain path in the hypoplastic constitutive equations. In this work, we will extend a simple yet powerful hypoplastic model for cohesive soil so that the extended model is able to account for two important features, namely cohesion and strain softening in cohesive soils.

The remaining part of this paper is arranged as follows. Section 2 is devoted to the presentation of the framework of hypoplasticity and a reference hypoplastic model for granular materials. In Sect. 3, a basic hypoplastic constitutive model and its explicit yield surface are briefly outlined. In addition, two important extensions of the basic model, namely the extensions for cohesion and strain softening are given in Sect. 4. In Sect. 5, the performance of this extended model is examined by performing two boundary value problems using finite element method. Finally, some concluding remarks are summarized in Sect. 6.

2 Framework of Hypoplasticity

Hypoplasticity is a nonlinear rate-form constitutive theory that has long been used for analysing mechanical behaviours [10] and other boundary value problems [11] in granular material. In the framework of hypoplasticity, the constitutive equation is written in two parts, representing respectively reversible and irreversible behaviors of soils.

$$\mathring{\boldsymbol{\sigma}} = \mathscr{L}(\boldsymbol{\sigma}, \dot{\boldsymbol{\varepsilon}}) + N(\boldsymbol{\sigma}, \dot{\boldsymbol{\varepsilon}}), \tag{1}$$

where $\boldsymbol{\sigma}$ is the Cauchy stress tensor, $\dot{\boldsymbol{\varepsilon}}$ is the stretching tensor, $\mathring{\boldsymbol{\sigma}}$ is the Jaumann rate of the Cauchy stress tensor defined in terms of the time-derivative of the Cauchy stress tensor $\dot{\boldsymbol{\sigma}}$ and the spin tensor $\boldsymbol{\omega}$

$$\mathring{\boldsymbol{\sigma}} = \dot{\boldsymbol{\sigma}} + \boldsymbol{\sigma} \cdot \boldsymbol{\omega} - \boldsymbol{\omega} \cdot \boldsymbol{\sigma}, \tag{2}$$

The term $\mathscr{L}$and N denote the linear and nonlinear components in tensor $\dot{\boldsymbol{\varepsilon}}$,which are isotropic tensor-valued functions consisted by the terms from the representation theorem. Here $\mathscr{L}(\boldsymbol{\sigma})$ is a fourth-order tensor.

We start with the formulation of Wu and Kolymbas [12] and write the hypoplastic rate-constitutive equation as the sum of linear and nonlinear terms of the strain rate $\dot{\boldsymbol{\varepsilon}}$

$$\mathring{\boldsymbol{\sigma}} = \mathscr{L}(\boldsymbol{\sigma}) : \dot{\boldsymbol{\varepsilon}} + \boldsymbol{N}(\boldsymbol{\sigma})\|\dot{\boldsymbol{\varepsilon}}\| \tag{3}$$

where $\|\dot{\boldsymbol{\varepsilon}}\| = \sqrt{\mathrm{tr}\dot{\boldsymbol{\varepsilon}}^2}$ stands for the Euclidean norm of the stretching tensor. The colon: denotes an inner product between two tensors.

2.1 A Reference Model

On the basic of above concepts, a simple hypoplastic constitutive equation for granular materials has been proposed by Wu and Bauer [13] to solve some boundary value problems. It is composed of two linear and two non-linear terms in stretching tensor $\dot{\boldsymbol{\varepsilon}}$.

$$\mathring{\boldsymbol{\sigma}} = C_1 \mathrm{tr}(\boldsymbol{\sigma})\dot{\boldsymbol{\varepsilon}} + C_2 \frac{\mathrm{tr}(\boldsymbol{\sigma}\dot{\boldsymbol{\varepsilon}})}{\mathrm{tr}\boldsymbol{\sigma}}\boldsymbol{\sigma} + (C_3 \frac{\boldsymbol{\sigma}^2}{\mathrm{tr}\boldsymbol{\sigma}} + C_4 \frac{\boldsymbol{\sigma}^{*2}}{\mathrm{tr}\boldsymbol{\sigma}})\|\dot{\boldsymbol{\varepsilon}}\| \tag{4}$$

where $C_i (i = 1, 2, 3, 4)$ are dimensionless parameters. The deviatoric stress tensor $\boldsymbol{\sigma}^*$ in the Eq. (4) is defined by $\boldsymbol{\sigma}^* = \boldsymbol{\sigma} - 1/3(\mathrm{tr}\boldsymbol{\sigma})\boldsymbol{\delta}_{ij}$, with $\boldsymbol{\delta}_{ij}$ being Kronecker delta.

The four parameters can be identified with a single triaxial compression test. Details of calibration procedure using the initial tangential stiffness E_i, the initial Poisson ratio υ_i, the internal friction angle φ and the dilatancy angle ψ to determine the material parameters can be found in the work by Wu and Bauer [13].

3 The Basic Hypoplastic Constitutive Model

Bauer [9] found out that constitutive Eq. (4) predicts critical state (defined by $\mathring{\boldsymbol{\sigma}} = 0$ and $\mathrm{tr}\dot{\boldsymbol{\varepsilon}} = 0$) for any paths only if the two parameters in the non-linear term of Eq. (4) satisfy:

$$C_3 = -C_4, \tag{5}$$

As a consequence of Eq. (5), the number of parameters in Eq. (4) reduces to three. This changing, however, restricts the predictive capability of this model. For instance, the initial Poisson ratio cannot be varied, which consequently results in another shortcoming: the volumetric strain cannot be changed in the critical state of a specimen with a initial void ratio near critical [1]. To resolve this problem, a new term $\mathrm{tr}(\dot{\boldsymbol{\varepsilon}})\boldsymbol{\sigma}$ introduced by Wang [4] is added to the constitutive Eq. (4), and thus the number of material parameters restored four. The modified constitutive equation can be rewritten out in full:

$$\mathring{\sigma} = C_1 \mathrm{tr}(\sigma)\dot{\varepsilon} + C_2 \mathrm{tr}(\dot{\varepsilon})\sigma + C_3 \frac{\mathrm{tr}(\sigma\dot{\varepsilon})}{\mathrm{tr}\sigma}\sigma + C_4(\sigma + \sigma^*)\|\dot{\varepsilon}\|, \tag{6}$$

Without causing confusion, the same notations for the four parameters are retained in the above equation. The new term $\mathrm{tr}(\dot{\varepsilon})$ maintains the critical state for all paths, since it vanishes in a critical state ($\mathrm{tr}\dot{\varepsilon} = 0$).

Some practical applications have been conducted based on the basic constitutive model. Recently, the micropolar theory has been successfully incorporated into this basic constitutive model to study the evolution of shear band in granular materials [14]. Also, it has been implemented into the smoothed particle hydrodynamics code (SPH) for debris flow materials [15].

3.1 Failure Surface and Flow Rule

The failure surface and flow rule of the above hypoplastic model is known as byproducts of its equation [13, 16]. Hence, the failure surface and flow rule of the constitutive Eq. (6) can be explicitly expressed using the flowing failure definition.

A material element is considered to be at failure if, for a given stress σ, there exits at least one strain rate $\dot{\varepsilon} \neq 0$ resulting in vanish stress rate. A straightforward way of the statement is given as follows, i.e

$$\mathring{\sigma} = \mathscr{L}(\sigma) : \dot{\varepsilon} - N(\sigma)\|\dot{\varepsilon}\| = 0 \tag{7}$$

The direction of strain rate at failure can be readily obtained from Eq. (7):

$$\frac{\dot{\varepsilon}}{\|\dot{\varepsilon}\|} = \mathscr{L}^{-1} : N \tag{8}$$

By making use of the fact that $(\dot{\varepsilon}/\|\dot{\varepsilon}\|) : (\dot{\varepsilon}/\|\dot{\varepsilon}\|) = 1$, the failure criterion can be derived:

$$f^s = N^{\mathrm{T}} : (\mathscr{L}^{\mathrm{T}})^{-1} : \mathscr{L}^{-1} : N - 1 = 0 \tag{9}$$

The explicit formula of the failure surface can be readily obtained using the symbolic computational program *Mathematica*, which gives rise to the failure surface in the following form:

$$f^s = \sqrt{J_2} - q_c I_1 = 0, \tag{10}$$

where J_2 and I_1 are respectively the second deviatoric stress invariant and the first stress invariant. q_c is a constant determined by the dimensionless parameters $C_i (i = 1, 2, 3, 4)$. Further parametric study reveals that the parameter q_c is only dependent

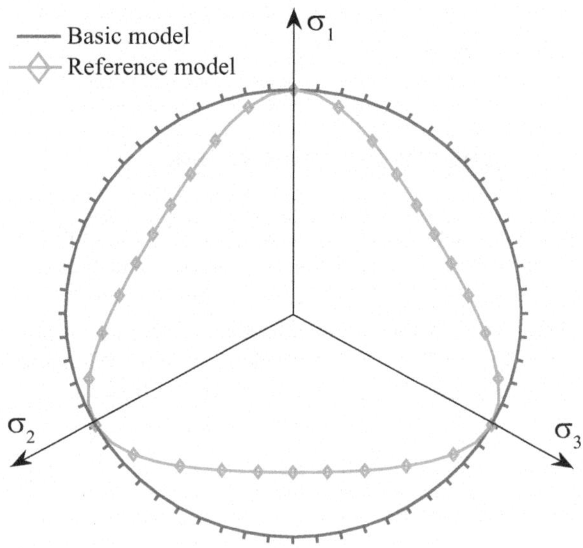

Fig. 1 Failure surface and flow rule on an octahedral plane. the failure surface of the basic model circumscribes that of the reference model on an octahedral plane. The radial red lines denote the direction of strain rate at failure

on the internal friction angle of materials, and it takes the formula as:

$$q_c = \frac{2sin\varphi}{\sqrt{3}(3 - sin\varphi)} \tag{11}$$

The failure surfaces of the reference model and the basic model shown in Fig. 1 indicate that the former one is a cone in the principal stress space, whereas the latter one is a circle enclosing the former one. Therefore, it possesses the same yield limit for both compression and extension. Equation (8) denotes the direction of the strain rate, that is flow rule in hypoplasticity. The flow direction at failure is also plotted in Fig. 1 with a short radial line. The direction of the strain rate is not usually normal to the failure surface, which suggests that the flow rule is non-associated.

Apparently, the Eq. (10) is coincident with the yield criterion of Drucker-Prager, which takes the following form [17]:

$$f^s = \sqrt{J_2} - q_\varphi I_1 - k_\varphi = 0, \tag{12}$$

where q_φ and k_φ are parameters determined by the internal friction angle φ and cohesion c of the soil materials. If we assume that the Drucker-Prager yield surface circumscribes the Mohr-Coulomb yield surface, the expressions for q_φ and k_φ are:

$$q_\varphi = \frac{2sin\varphi}{\sqrt{3}(3 - sin\varphi)}, \quad k_\varphi = \frac{6c\ cos\varphi}{\sqrt{3}(3 - sin\varphi)}, \tag{13}$$

Noting that q_c is identically equal to q_φ in this case, and the Eq. (11) reduces Eq. (10) if cohesion $c = 0$.

4 Extensions to Cohesive Soil

Both the constitutive Eqs. (4) and (6), albeit their mathematical simplicity, are capable of capturing the salient features of granular materials, such as nonlinearity, failure and dilatancy [4, 13]. However, some behaviors such as cohesion, strain softening, effects of void ratio, and inherent anisotropy cannot be taken into account. In the following subsections, we suggested some extensions concerning the cohesion and strain softening of cohesive soils for the basic model.

4.1 Cohesion

As discussed above, the limit state of Eq. (6) leads to a conical surface with its apex in the conical origin of the principal stress space. Therefore, the Eq. (6) is not able to account for tensile stress. Inspired by the similarity between the failure surface of the basic model and Drucker-Prager model. The parameter k_φ can be introduced into Eq. (10) to account for cohesion, then the yield surface reads:

$$f^s = \sqrt{J_2} - q_c I_1 - k_\varphi = 0 \tag{14}$$

For the constitutive Eq. (16), the same effect can be achieved by simply replacing the stress tensor $\boldsymbol{\sigma}$ with the following translated stress tensor [4]:

$$\boldsymbol{\sigma_c} = \boldsymbol{\sigma} - p_c \mathbf{1}, \tag{15}$$

where p_c is related to cohesion c and friction angle φ of Mohr-Coulomb yield criterion. With this approach, the apex of the circle cone is translated to point $s(-p_c, -p_c, -p_c)$ in the principal stress space. Substitution of the stress point $s(-p_c, -p_c, -p_c)$ into Eq. (14) yields the magnitude of $p_c = c\ /tan\varphi$. Thus, the constitutive Eq. (6) can be extended as follows:

$$\mathring{\boldsymbol{\sigma}} = C_1 \mathrm{tr}(\boldsymbol{\sigma}_c)\dot{\boldsymbol{\varepsilon}} + C_2 \mathrm{tr}(\dot{\boldsymbol{\varepsilon}})\boldsymbol{\sigma}_c + C_3 \frac{\mathrm{tr}(\boldsymbol{\sigma}_c \dot{\boldsymbol{\varepsilon}})}{\mathrm{tr}\boldsymbol{\sigma}_c}\boldsymbol{\sigma}_c + C_4(\boldsymbol{\sigma}_c + \boldsymbol{\sigma}_c^*)\|\dot{\boldsymbol{\varepsilon}}\| \tag{16}$$

The effect of cohesion is demonstrated using two numerical triaxial compression tests on a cohesionless and cohesive soils. The material parameters presented in Table 1 for Eq. (16) are used for the numerical simulations. The cohesion is $c =$ 10 kPa and the confining pressure is $\sigma_3 =$ 100 kPa. The result presented in Fig. 2

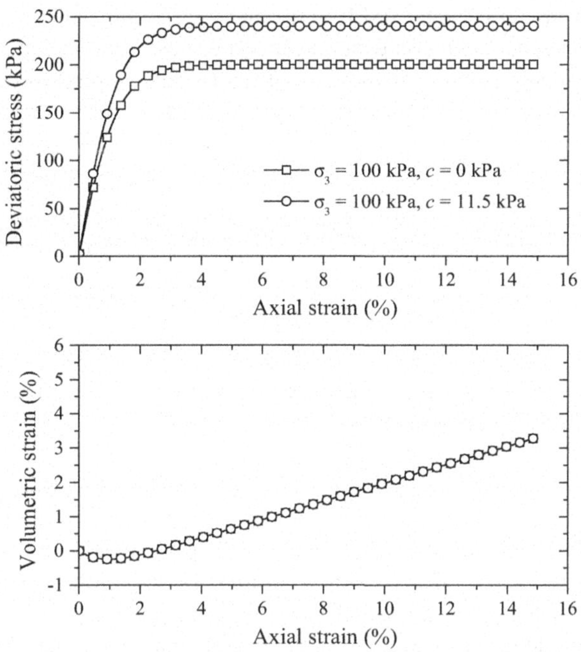

Fig. 2 Numerical simulation of triaxial compression test with confining pressure $\sigma_3 = 100\,\text{kPa}$, and cohesion $c = 0/11.5\,\text{kPa}$

reveals that the cohesion can influence the strain- stress response, that is, increase the deviatoric stress, while it cannot influence the volumetric response.

4.2 *Strain Softening*

The stress-strain response of the constitutive Eq. (6) is governed by the interaction between the linear and non-linear terms. The linear term can be viewed as constructive, whereas the non-linear term can be considered as destructive. Any perturbation in either the linear part or the non-linear part at failure will bring the constitutive equation away from its equilibrium position and give rise to either hardening or softening [14, 18].

To account for the softening behaviour in the stress-strain response, we can simply enhance the non-linear term causing the constitutive response to be contractive. This

Table 1 Parameters of the triaxial compression tests

Parameters	E_i (MPa)	υ_i	φ	ψ	c(kPa)	C_1	C_2	C_3	C_4
Value	170	0.2	30°	15°	11.5	−47.2	−81.36	−692.84	−153.05

Note: The parameter constants $C_i(1, 2, 3, 4)$ are identified with the material parameters E_i, υ_i, φ and ψ

aim can be achieved by introducing a strain-path-related factor into the non-linear term as a perturbation that incorporates the effect of strain-path on the constitutive response. In the framework of Eq. (3), the general form of the extended model reads:

$$\mathring{\boldsymbol{\sigma}} = \mathscr{L}(\boldsymbol{\sigma}_c) : \dot{\boldsymbol{\varepsilon}} - f_d \boldsymbol{N}(\boldsymbol{\sigma}_c)\|\dot{\boldsymbol{\varepsilon}}\|, \tag{17}$$

and thus the complete form of the modified equation is obtained

$$\mathring{\boldsymbol{\sigma}} = C_1 \mathrm{tr}(\boldsymbol{\sigma}_c)\dot{\boldsymbol{\varepsilon}} + C_2 \mathrm{tr}(\dot{\boldsymbol{\varepsilon}})\boldsymbol{\sigma}_c + C_3 \frac{\mathrm{tr}(\boldsymbol{\sigma}_c \dot{\boldsymbol{\varepsilon}})}{\mathrm{tr}\boldsymbol{\sigma}_c}\boldsymbol{\sigma}_c + f_d C_4(\boldsymbol{\sigma}_c + \boldsymbol{\sigma}_c^*)\|\dot{\boldsymbol{\varepsilon}}\|, \tag{18}$$

Here, f_d is a scale factor, which represents the effect of strain path. In the present study f_d is assumed to be:

$$f_d = (1-\alpha)e^{-\beta l} + \alpha, \tag{19}$$

in which the constant α relates to the ratio of peak stress to residual stress in the strain-stress curve, thus $\alpha \geqslant 1$ for strain softening and the scale factor $f_d = 1$ if $\alpha = 1$. The constant β is a positive parameter that relates to the speed of strain softening. l is the length of strain path [19], which can be updated during the deformation process according to the following evolution equation:

$$l = \int_{t_f}^{t} \|\dot{\boldsymbol{\varepsilon}}(t)\| d\tau. \tag{20}$$

where t_f denotes the time when failure occurred in the material. This means the scale factor will be invoked if failure has occurred in the soils.

For a better understanding of the evolution function, the relation between strain path l and scalar factor f_d is presented in Fig. 3. It shows that the increase rate of scale factor f_d is positively related to the constant β, while the maximum value of f_d only depends on α. Therefore, the value of scalar factor f_d lies within the range $(1, \alpha)$, which can give rise to enhancement of the non-linear term of constitutive Eq. (6) and cause degradation of stress over strain history.

In order to demonstrate the influence of the scale factor f_d on the strain-stress response of Eq. (18), a numerical triaxial compression test on cohesionless soil with a constant confining pressure 100 kPa is conducted. The parameters shown in Table 1 and different constants for Eq. (19) are used in this simulation and the simulation results are presented in Fig. 4. The essential effects of the scale factor f_d illustrated above seem to have been produced by the extended hypoplastic constitutive Eq. (18). Unlike a constant stress remained in the post-failure condition in Fig. 2, a peak can be observed in the strain-stress curve. After the peak, there is a gradual reduction of the stress over strain history and a larger value of α can give rise to a more pronounced reduction of the stress. Eventually, the stress reduces to a constant value that is controlled by the parameter α. The volumetric strain response can be affected by the

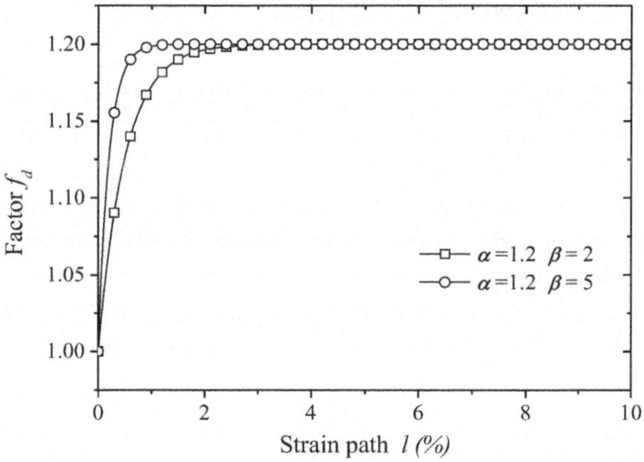

Fig. 3 Relation between strain path l and scalar factor $f_d = (1 - \alpha)e^{-\beta l} + \alpha$, which destructs the non-linear part of the model and gives rise to reduction of the stress rate. α relates the ratio of peak strength to residual strength, and β denotes the speed of reduction of stress rate after failure

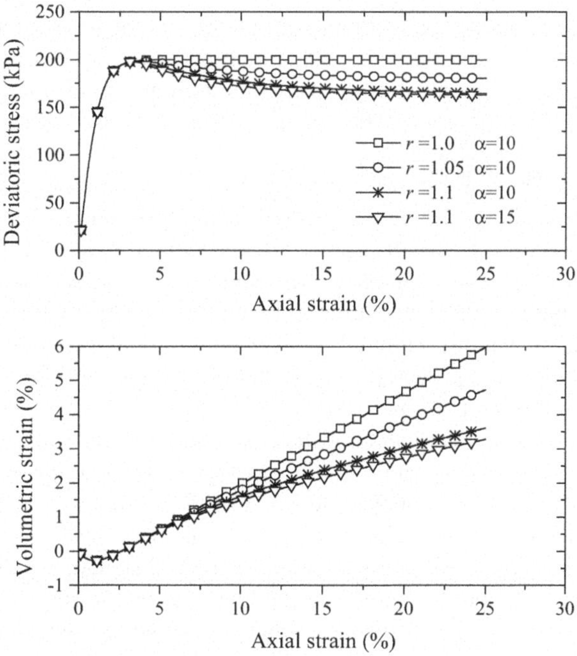

Fig. 4 The influence of the scale factor f_d on the strain-stress response of triaxial compression test

scale factor as well. As illustrated in Fig. 4, the volumetric strain rate decreases with the constant α and β. The possible effect of parameter β on the stress-strain response has been examined as well. As shown in Fig. 4, β can only influence the speed of stress reduction in post-failure condition but it cannot change the final stress.

5 Numerical Examples

The extended constitutive model has been implemented into a commercial FEM code Abaqus to solve some boundary value problems. In the following section, two numerical examples are presented with using the extended hypoplastic model. In the first example, a tunnel excavation in a homogeneous soil is simulated. Different values of cohesion for the excavated soil are used in this simulation. We also conduct a plane strain test to investigate the effect of the scale factor. Noting that there is no attempt to match the results to the practical process in tunnelling simulation. Instead, we only focus on the performance of the extensions.

5.1 Excavation in Soil

The model has a dimension of 60 × 60 m with a tunnel (8 m in diameter) being excavated in the soil. The numerical model is defined using Abaqus with 446 plate strain elements. For the sake of simplicity, an asymmetric element is chosen for the FEM simulation. The material parameters used in this simulation are shown in the following Table 2.

There are two steps to carry out the tunnel excavation: a geostatic step and a tunnelling step. In the geostatic step, the boundary condition for the bottom nodes is given by fixing the vertical displacement. Nodes on lateral boundaries are fixed in the horizontal direction. The moment on all nodes is set to be zero. Gravity load is applied in this step after all boundary conditions are applied. In the second step, the elements of the tunnel is removed to model the excavation process. In the practical tunnelling process, however, a tunnel lining will be set up, which is not considered in this simulation.

The contour plots of the horizontal and vertical displacements induced by excavation are presented in Fig. 5, which reveal that the majority surface settlements are located several meters deviated from the tunnel centre line. Fig. 6 shows the surface settlement on the ground surface calculated using different values of cohesion and the largest surface settlement is 10 m distance from the tunnel centre line. Corresponding deformations at the final excavation stage along the vertical cross-section 10 m distance from the tunnel centreline are depicted in Fig. 7 Differences in the settlement magnitudes are due to different responses of cohesion in the model. A remarkable rebound near the excavation can be noticed, which agrees with the practical tunnelling. Both figures reveal that a larger value of cohesion results in a smaller surface settlement, so the effect of cohesion in the extended model is produced.

Table 2 Parameters of the excavated soil

Parameters	E_i (MPa)	υ_i	φ	ψ	c(kPa)	α	β
Value	220	0.2	30°	10°	50/100	1	10

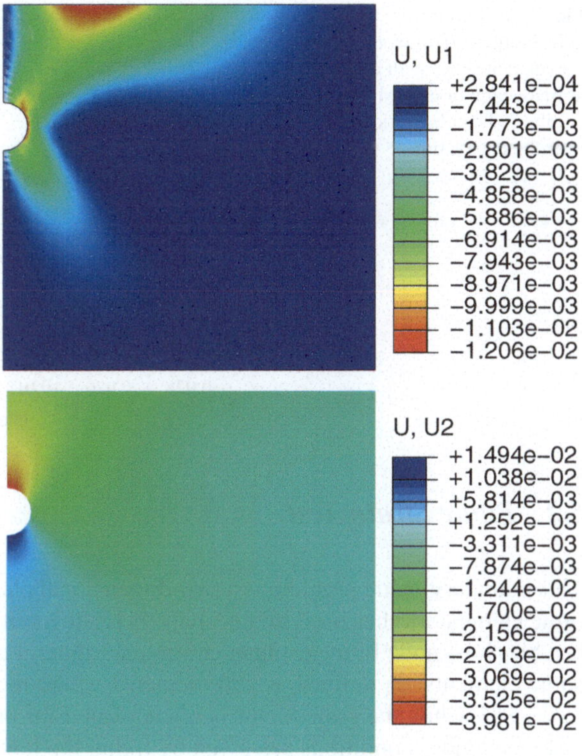

Fig. 5 Contour plots of horizontal displacement (U1) and vertical displacement (U2)

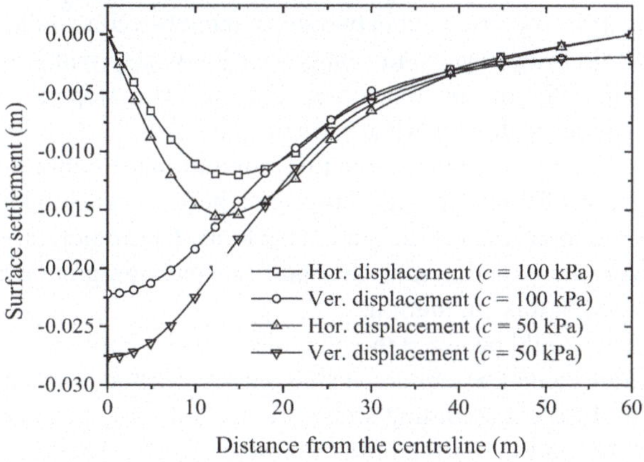

Fig. 6 Surface settlement above tunnel centreline. Two different cohesion parameters, 50 and 100 kPa, are used in this simulation, resulting in different surface settlements

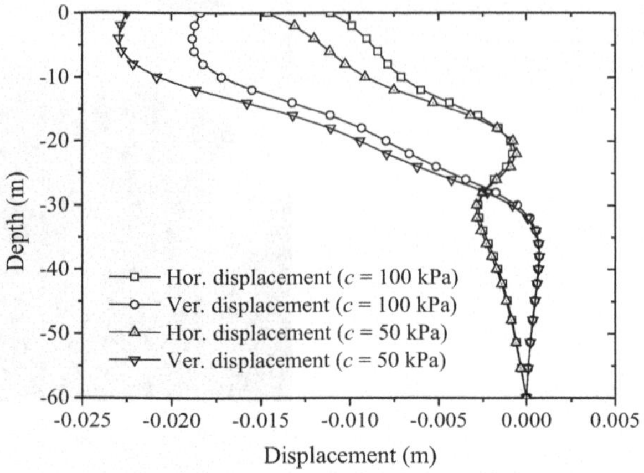

Fig. 7 Variation of deformations 10 m from tunnel centreline with depth at final stage of excavation. A rebound can be noticed near the excavation

5.2 *Plane Strain Test*

This boundary-value problem is a two-dimensional plane strain test. The model with finite element meshes used in this analysis is shown in Fig. 8. This numerical model has 880 four-node bilinear plane strain quadrilateral elements. As shown in Fig. 8, some elements are defined as soft elements where the scale factor will be invoked ($\alpha = 1.5$), while the scale factor is not activated for the rest elements with $\alpha = 1.0$. The boundary conditions are given by fixing the bottom of the model. This test is simulated by applying a uniform downward displacement on the top surface. The analysis is carried out in two steps: namely a consolidation step and a shearing step. In the first step, a geostatic analysis is invoked to make sure that equilibrium is satisfied within the soil specimen. Step 2 is a shearing step in which the loading plate is forced to displace downward at a small rate.

The parameters used in this simulation are calibrated with a triaxial compression test of silty sand. Figure 9 presents the soil's triaxial behaviour under three confining pressures (102.4, 172.4 and 310.3 kPa). The numerical stress-strain result is compared with the experimental data and reasonable agreement of stress-strain response is noted in this simulation.

It should be noted that the dilative behaviour in the volumetric strain versus axial strain cannot be captured using the same initial Poisson ratio. Different Poisson ratios for different confining stress are needed for capturing the dilative phenomenon. As shown in Fig. 10, the initial Poisson ration is 0.3, 0.33 and 0.2 for confining pressure of 102.4, 172.4 and 310.3 kPa, respectively. In this simulation, therefore, we use 0.3 as the Poisson ratio to identify the hypoplastic parameters. Of particular interest is the strain softening behaviour in the post-failure stage, where the numerical results perfectly captured this important phenomenon by invoking the scale factor in the updated model. This implies that the implementation of the extended model in the finite element program used herein is done correctly.

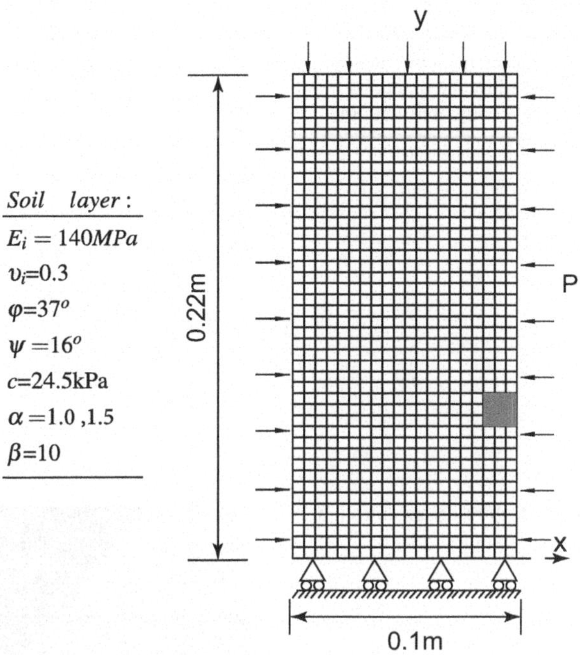

Fig. 8 Parameters and plane strain test configuration

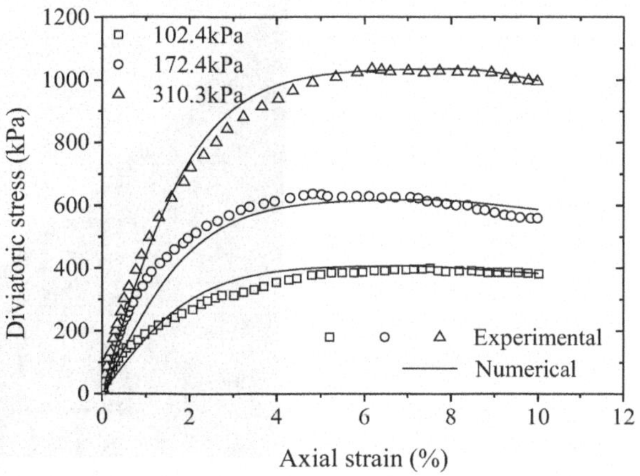

Fig. 9 Stress-strain behaviour of a dense silty sand: result predicted by the extended model is compared with the experimental data

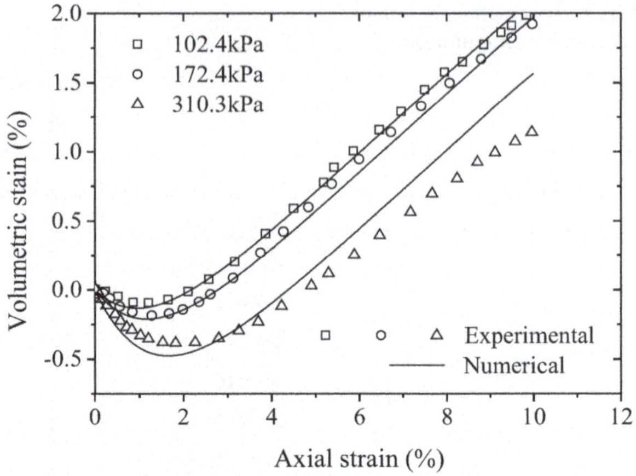

Fig. 10 Volumeric strain-axial strain behavior of a dense silty sand: result predicted by the extended model is compared with the experimental data

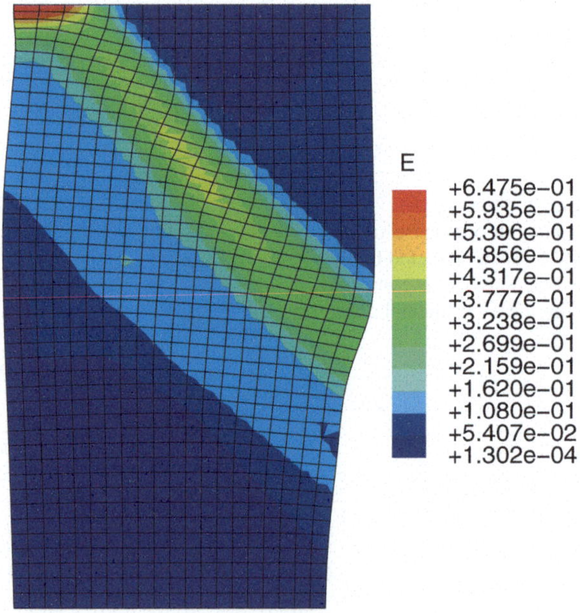

Fig. 11 Strain concentration along the failure plane

Figure 11 presents the distribution of the strain at an advanced stage of loading. It is known that strain localisation cannot occur in a perfectly homogeneous FEM model. However, owing to the strain softening that occurred in the soft elements, a shear concentration phenomenon can be observed. With the increase of axial displacement, the concentrated zone results in a failure plane with an angle of 49° and further leads to failure of this test. In addition, the distortion within this concentrated strain zone (also known as a shear band) is evident, and the zones above and below the shear band have a minimal amount of distortion.

6 Conclusions

The presented paper discusses the extensions of a simple hypoplastic model to describe cohesive soils. The main conclusions of this study are drawn as follow:

(1) The explicit yield function of the basic hypoplastic model is derived, which is of Drucker-Prager type with its apex in the conical origin, and circumscribes the Mohr-Coulomb yield surface in the principal stress space.
(2) To enhance the capacity of the basic model, a translated tensor and a scale function are introduced. Combining these two extensions, the proposed model is able to describe the effect of cohesion and strain softening behaviour of cohesive soils.
(3) The extended constitutive model is implemented in a finite element code to solve some boundary value problems. FEM numerical studies of a tunnel excavation and a plane strain test are carried out. The results reveal that the extensions for cohesion and strain softening have been produced.

Acknowledgements The research carried out in this paper is partly funded by the project "GEO-RAMP" within the RISE programme of Horizon 2020 under grant number 645665. The first author is grateful to the financial support from the Otto Pregl Foundation for Fundamental Geotechnical Research in Vienna.

References

1. Wu, W., Kolymbas, D.: Hypoplasticity then and now. In: Constitutive modelling of granular materials, pp. 57–105. Springer, (2000)
2. Bauer, E., Wu, W.: A hypoplastic constitutive model for cohesive powders. Powder Technol. **85**, 1–9 (1995)
3. Weifner, T., Kolymbas, D.: A hypoplastic model for clay and sand. Acta Geotechnica **2**, 103–112 (2007)
4. Wang, X., Wu, W.: An updated hypoplastic constitutive model, its implementation and application. In: Bifurcations, Instabilities and Degradations in Geomaterials, pp. 133–143. Springer (2011)
5. Mašín, D.: A hypoplastic constitutive model for clays. Int. J. Numer. Anal. Methods Geomech. **29**, 311–336 (2005)
6. Niemunis, A.: Extended hypoplastic models for soils, Inst. für Grundbau und Bodenmechanik (2003)
7. Huang, W.X., Wu, W., Sun, D.A., et al.: A simple hypoplastic model for normally consolidated clay. Acta Geotechnica **1**, 15–27 (2006)
8. Gudehus, G.: A comprehensive constitutive equation for granular materials. TASK Quarterly: scientific bulletin of Academic Computer Centre in Gdansk **4**, 319–342 (2000)

9. Bauer, E.: Calibration of a comprehensive hypoplastic model for granular materials. Soils Found. **36**, 13–26 (1996)
10. Herle, I., Kolymbas, D.: Hypoplasticity for soils with low friction angles. Comput. Geotechnics **31**, 365–373 (2004)
11. Hleibieh, J., Wegener, D., Herle, I.: Numerical simulation of a tunnel surrounded by sand under earthquake using a hypoplastic model. Acta Geotechnica **9**, 631–640 (2014)
12. Wu, W., Kolymbas, D.: Numerical testing of the stability criterion for hypoplastic constitutive equations. Mech. Mater. **9**, 245–253 (1990)
13. Wu, W., Bauer, E.: A simple hypoplastic constitutive model for sand. Int. J. Numer. Anal. Methods Geomech. **18**, 833–862 (1994)
14. Lin, J., Wu, W., Borja, R.I.: Micropolar hypoplasticity for persistent shear band in heterogeneous granular materials. Comput. Methods Appl. Mech. Eng. **289**, 24–43 (2015)
15. Peng, C., Wu, W., Yu, H.S., et al.: A sph approach for large deformation analysis with hypoplastic constitutive model. Acta Geotechnica **10**, 703–717 (2015)
16. Wu, W., Niemunis, A.: Failure criterion, flow rule and dissipation function derived from hypoplasticity. Mech. Cohesive-Frictional Mater. **1**, 145–163 (1996)
17. Yu, T., Teng, J.G., Wong, Y.L., et al.: Finite element modeling of confined concrete-I: Drucker-Prager type plasticity model. Eng. Struct. **32**, 665–679 (2010)
18. Wu, W., Bauer, E., Kolymbas, D.: Hypoplastic constitutive model with critical state for granular materials. Mech. Mater. **23**, 45–69 (1996)
19. Wu, W., Bauer, E., Niemunis, A., et al.: Visco-hypoplastic models for cohesive soils. Mod. Approaches to Plasticity, pp. 365–383 (1993)

Temperature Effect on the Compressive Strength of Frozen Soils: A Review

Guofang Xu, Jilin Qi and Wei Wu

Abstract Among the numerous influencing factors on the strength behavior of frozen soil, such as grain size, temperature, water content, strain rate, and confining pressure, etc. the temperature is a crucial one. In this paper a number of references concerning the temperature effect are reviewed. It is found that the strength of different frozen soils increases with the decrease of temperature below the freezing point of water, and the quantitative dependence of the strength on temperature can be primarily described by two kinds of functions, namely power function and linear function. The function parameters, including the exponent and slope, are collected. Moreover, how the other influencing factors affect the temperature parameter is also analyzed. It is found that, for example, the decrease in strain rate for testing or the increase in dry density of specimen will cause an increase in the exponent in the power function of temperature. Besides the temperature often encountered in engineering problem, the strength behavior of frozen soil at cryogenic temperatures is also reviewed.

G. Xu (✉)
State Key Laboratory of Geomechanics and Geotechnical Engineering, Institute of Rock and Soil Mechanics, Chinese Academy of Sciences, Xiaohongshanstreet 2, Wuhan 430071, China
e-mail: gfxu@whrsm.ac.cn

J. Qi
School of Civil and Transportation Engineering, Beijing University of Civil Engineering and Architecture, Zhanlanguanstreet 1, Beijing 100044, China
e-mail: jilinqi@bucea.edu.cn

W. Wu
Institute of Geotechnical Engineering, University of Natural Resources and Life Sciences, Vienna, Feistmantelstrasse 4, 1180 Vienna, Austria
e-mail: wei.wu@boku.ac.at

W. Wu (ed.), *Recent Advances in Geotechnical Research*,
Springer Series in Geomechanics and Geoengineering,
https://doi.org/10.1007/978-3-319-89671-7_19

1 Introduction

Frozen soil is a complex multiphase material. Generally it contains soil grain, ice, air, and unfrozen water. The most important characteristic which differs from other similar materials–such as unfrozen soils and the majority of artificial composites–is that under natural conditions its matrix, which is composed mostly of ice and water, changes continuously with varying temperature and applied stress [1]. There is only one type of ice present in intergranular pores, i.e. hexagonal ice, the most common form of ice, but the unfrozen water exists in two states: strongly bound and weakly bound water [1]. The former is the water film surrounding mineral particles and adheres to them by high intermolecular forces that suppress freezing, even at very low temperatures [3]. The latter is weakly bounded and can migrate freely and be frozen more easily.

Owing to the components of ice and unfrozen water, the strength of frozen soil is significantly affected by temperature. When the temperature decreases, part of the unfrozen water will be frozen and the cementation of ice will be enhanced, which will cause a considerable increase of the cohesion, pursuantly the strength of frozen soil. In this paper, a number of references concerning the temperature effect are reviewed. It is found that the relationship between the strength and temperature can be described by two kinds of simple functions, namely power function and linear function. Compared to the attempts in the past to express the strength of frozen soils with temperature by means of the theory of rate process, the simple functions give a better fit to experimental results. The parameters in the simple functions of temperature are collected and the effects of other influencing factors, such as grain size, water content, strain rate, and confining pressure, etc. on the temperature parameter are also analyzed.

2 Temperature Effect on the Strength of Frozen Soils

A number of compression tests at different temperatures and strain rates are conducted on various frozen soils by investigators [5–11, 13, 14, 16, 17, 19–21]. Based on the test results, the temperature effect on the strength behaviors of frozen soils is studied. In the following the information and results of the tests are presented.

Uniaxial compression tests at fast loading rates (50–90 MPa/min) were conducted by Tsytovich [17] on different types of frozen soils, then Tsytovich proposed the following equation for the strength of frozen soils

$$\sigma = a + b|\theta|^n \tag{1}$$

in which a, b, and n are parameters; θ is temperature. Tsytovich pointed out that the equation was applicable when the temperature is not lower than 15 °C. Based on the test results of naturally compact sandy loam and clay, Shusherina and Vyalov [16] suggested a value of 1.0 for n, i.e. the strength of the frozen soils depended linearly on temperature. However, according to the test results in the Igarka Scientific Center of

Russian Academy of Sciences, n has a value close to 0.5, i.e. the strength is related to temperature by a power function. The power relationship was also presented by Hu et al. [11] for silty clay and sandy silt from −5 to −20 °C after one or two freezing-thaw cycles.

Zhu and Carbee [21] conducted uniaxial compressive strength tests on remolded, saturated frozen Fairbanks silt under six constant machine speeds (from 6.2×10^{-2} to $1.1 \times 10^{-6}\,s^{-1}$), seven temperatures from −0.5 to −10 °C, and three dry densities. The soil tested in the investigation had the following physical properties: plastic limit of 34.2%, liquid limit of 38.4%, organic content of 5.49%, and a specific gravity of 2.68. The test results for various applied strain rates were presented in a plot of log σ_m versus $\log(\theta/\theta_0)$, where θ is the test temperature and θ_0 is a reference temperature taken as −1 °C. It is shown that the peak strength of the frozen silt significantly increased with decreasing temperature. The strength equation can be written as

$$\sigma_m = A(\theta/\theta_0)^m \tag{2}$$

where A is an empirical parameter with the dimension of stress, and m is a dimensionless parameter. The same relation was also reported by Wijeweera and Joshi [19] for saturated fine-grained frozen soils in uniaxial compression tests at temperatures between −5 and −17 °C under a constant strain rate of $0.01\,s^{-1}$; and also reported by Li et al. [13] for frozen Lanzhou loess in uniaxial compression tests from −2 to −15 °C under the strain rates 6×10^{-4} to $5 \times 10^{-7}\,s^{-1}$ for the dense samples ($1.76\,g/cm^3$) and 7×10^{-4} to $1 \times 10^{-6}\,s^{-1}$ for the loose samples ($1.58\,g/cm^3$). The values of A and m, which were obtained by regression analysis, for various strain rates are listed in Tables 1 and 2.

Cylindrical samples of frozen Ottawa sand with diameter of 50.8 mm and length of about 140 mm were prepared by Parameswaran [14] in a split Plexiglas mould by a method described by Baker [4]. Then some uniaxial unconfined compression tests were carried out on the saturated frozen sand containing about 20% by weight of water, at temperatures between −2 and −15 °C and strain rates varying between 10^{-7} and $10^{-2}\,s^{-1}$. The test results for the strength against temperature at various strain rates were presented in a log-log plot. At relatively high strain rates, a linear variation can be found and the slope of the lines is about 0.44. This suggests that strength varies with temperature according to the following equation

Table 1 Values of A and m in Eq. (2) from Zhu and Carbee [21]

$\dot{\varepsilon}$ [s^{-1}]	A [MPa]	m	Correlation R^2
6.2×10^{-2}	5.27	0.49	0.9931
5.7×10^{-3}	2.84	0.49	0.9931
1.1×10^{-3}	1.87	0.49	0.9908
1.1×10^{-4}	1.00	0.59	0.9970
1.1×10^{-5}	0.59	0.73	0.9976
1.1×10^{-6}	0.44	0.84	0.9981

$$\sigma \propto \theta^s \tag{3}$$

where $s = 0.44$.

In the tests by Bragg and Andersland [7], the Wedron silica sand was adopted as the study material. The sand consists of sub-angular quartz particles with a uniform gradation (size range of 105–595 μm) and a coefficient of uniformity equal to 1.50. All samples were prepared in split aluminum molds with a sand volume fraction of 64%. Uniaxial compression tests at −6 °C and the strain rate of 1.0×10^{-7} s^{-1} to 1.0×10^{-5} s^{-1} were conducted on cylindrical samples to show the effects of temperature on the material properties of the frozen sand. The compression strength is observed to increase with decreasing temperatures at constant strain rates. Then Bragg and Andersland [7] proposed the same power expression as Parameswaran [14] to relate the compressive strength with temperature. For temperatures colder than −6 °C, the exponent s has the value of 0.49. For small temperature intervals, Bragg and Andersland [7] took the following relationship for the uniaxial compressive strength suggested by Andersland et al. [2] and Ladanyi [12].

$$\sigma_f \simeq \sigma_{c0} f(\theta) \tag{4}$$

where σ_{c0} is the strength (proof stress) for a temperature close to 0° and $f(\theta)$ can be approximated by

$$f(\theta) = 1 + \frac{\theta}{\theta_0} \tag{5}$$

Using a plot of strength versus temperature, θ_0 will be the intercept on the temperature axis and σ_{c0} will be the intercept on the strength axis. Equations (4) and (5) show a linear relationship between the strength and temperature. The linear relationship is also obtained by the following researchers.

Yamamoto and Springman [20] carried out a series of axial compression tests at constant strain rate on artificially frozen silty gravel at the temperatures between

Table 2 Values of A and m in Eq. (2) from Li et al. [13]

Dry density	Strain rate (s^{-1})	A [MPa]	m
1.76 g·cm^{-3}	6.50×10^{-4}	1.53	0.6992
	9.99×10^{-5}	1.37	0.6971
	8.42×10^{-6}	1.05	0.7396
	1.05×10^{-6}	1.01	0.7433
	5.14×10^{-7}	0.95	0.7455
1.58 g·cm^{-3}	7.32×10^{-4}	1.41	0.6690
	1.11×10^{-4}	1.00	0.7306
	9.52×10^{-6}	0.80	0.7274
	1.11×10^{-6}	0.68	0.7185

−3.0 and −0.3 °C. The specimens used in the tests had a volumetric ice content of $w_i = 76 - 84\%$, and the specific density of 2.71–2.72 g/cm^3. In the tests, Acoustic emissions were measured using a wide-band piezoelectric sensor to understand the mechanisms of the deformation behavior and micro-structural effects. The test results showed that an increase in temperature close to 0 °C resulted in a decrease in both the peak and residual deviatoric stresses. The temperature dependence of the peak and residual deviatoric stresses can be expressed by a linear regression for each strain rate as follows:

$$q = a + bT \tag{6}$$

where q is the deviatoric stress (kPa), T is the temperature (°C), a is the intercept (kPa), b is the coefficient (kPa/ °C). The values of a and b at different strain rates are given in Table 3.

The linear relationship was also reported by Chen et al. [8] for remoulded Shanghai soft clay with water contents from 29.4 to 35.7%. The temperatures and strain rates adopted in Chen's tests are [−15, −2] °C and [8×10^{-5}, 6×10^{-3}] s^{-1}, respectively. Hu et al. [10] also obtained linear relationships for different kinds of frozen soils, such as marine deposit, alluvium, and extremely weak granite, within the temperature range of −8 to −20 °C.

Haynes and Karalius [9] conducted a series of tests in uniaxial compression and tension to determine the effect of temperature on the strength of frozen Fairbanks silt. The test temperature ranged from 0 to −56.7 °C. Two machine speeds, 4.23 and 0.0423 cm/sec, were used for the constant displacement rate tests. From the highest to the lowest temperature, the compressive strength increased up to about one order of magnitude. Using the least squares method, relationships between the compressive strength and temperature were drawn for the two machine speeds employed in the tests.

$$\sigma_{100} = 7.64 - 1.362T \tag{7}$$

where σ is the strength in MN/m^2, and T is temperature in °C, the subscript 100 refers to the machine speed in inches per minute. The correlation of strength with temperature for the 0.0423 cm/sec machine speed was also developed with the method of least squares. It was found that the second-degree relationship was the best fit to the test results for the temperature range used in the tests. The equation obtained is

$$\sigma_1 = 2.15 - 0.33T + 0.01T^2 \tag{8}$$

Table 3 Values of a and b in Eq. (6) from Yamamoto and Springman [20]

Strain rate (s^{-1})	a [kPa]	b [kPa/ °C]	Correlation R^2
5.0×10^{-5}	1259.7	–626.5	0.99
5.0×10^{-6}	649.9	–364.7	0.78
1.0×10^{-6}	448.5	–285.0	0.97

where the symbols are the same as used above.

The strength behaviors at cryogenic temperatures are also studied by some investigators, such as Bourbonnais and Ladanyi [5, 6], and Sayles [15]. Uniaxial compression tests were conducted on Le Sueur sand by Bourbonnais and Ladanyi [6]. In the tests, the temperature range was [−6.7, −160] °C, the axial strain rate varies from 3×10^{-3} to 3×10^{-5} s^{-1}. The water content of the specimens is 20.5%, and the saturation is 92%. It is found that over such a wide temperature range the strength of the frozen sand varies in the pattern of a parabolic function of temperature, with the peak strength at about −100 °C. Similar tests were also conducted by Bourbonnais and Ladanyi [5] on over-consolidated Boom clay with a water content of 25%. The temperature used in the tests ranges from −16 to −154.4 °C, the strain rates from 2.6×10^{-3} to 9.2×10^{-6} s^{-1}. The test results showed that within the temperature range of [−16, −100] °C the strength of the frozen clay increased in a power function, while a linear function at the temperatures lower than −100 °C.

In the tests conducted on sandy silt, Ottawa sand (grain size 0.2–0.6 mm), and clay by Sayles [15], the temperature was as low as −180 °C. The test results showed that the strength of the frozen clay increased in a power function manner with the linear decrease in temperature. For the coarser-grained frozen materials such as sandy silt, the strength is found to increase perfectly linearly. For the frozen Ottawa sand, the observed strength increased parabolically with temperature decrease to −60 °C, then levelled off at the temperatures lower than −60 °C. Such complex variation of strength could be attributed to the wide range of temperature and can also be seen in Vialov's work [18].

3 Concluding Remarks

This paper reviews the variation laws of the compressive strength of frozen soil over different temperature ranges. Most of the laws are obtained for the temperature higher than −30 °C, which is often encountered in practical engineering. Only a few are with reference to cryogenic temperature, although they have less significance in engineering problems.

It is found that when the temperature is not lower than −30 °C, the compressive strength of frozen soil can be described by a power function (the linear functions obtained by some researchers are deemed power functions with the exponent (1). The exponent of the power function varies upon the material used and test condition. For example, $0.49 \leq n \leq 0.84$ is obtained by Zhu and Carbee [21] for saturated frozen Fairbanks silt at strain rate from 1.1×10^{-6} to 6.2×10^{-2} s^{-1} within the temperature range from −0.5 to −10 °C. The exponents obtained in all the reviewed references are summarized, as shown in Table 4. By comparing the materials and test conditions, the following conclusions on the magnitude of the exponent in the power functions, and the effects of the other influencing factors such as grain size, strain rate, etc. on the exponent are obtained.

Table 4 Summary of the temperature effect on the compressive strength of frozen soils

Researcher	Specimen	Temperature	Loading/Strain rate	Strength equation	Results
Tsytovich [17]	Silty sandy loam, $w = 21.3 - 23.1\%$, 0.005–0.05 mm weight content 61.2%	−0.5 to −10.3 °C	50–90 MPa/min	$\sigma = a + b\|\theta\|^n$	Igarka Scientific Center of RAS, $n \approx 0.5$; naturally compact sandy loam and clay, $n = 1.0$ [16]
Hu et al. [11]	Silty clay: $w = 23.52 - 24.07\%$, $\rho = 1.90 - 1.94\ \text{g/cm}^3$; sandy silt: $w = 28.30 - 29.03\%$, $\rho = 1.81 - 1.84\ \text{g/cm}^3$	−5 to −20 °C	Specimens damaged in 25–35 s	$\sigma = a + b\,(T)^n$	$n = 1$; silty clay: $a = [-1.37,\ 0.62]$, $b = [-0.37,\ -0.21]$; sandy silt: $a = [-1.4,\ 0.91]$, $b = [-0.39,\ -0.24]$
Zhu and Carbee [21]	Saturated frozen Fairbanks silt; PL = 34.2%, LL = 38.4%, SG = 2.68	−0.5 to −10 °C	6.2×10^{-2} to $1.1 \times 10^{-6}\ \text{s}^{-1}$	$\sigma_m = A(\theta/\theta_0)^m$	$A = [0.44,\ 5.27]$ MPa, $m = [0.49,\ 0.84]$
Wijeweera and Joshi [19]	Six fine-grained soils, Kaolin clay as an example. $w = 31.7 - 45\%$, $\gamma_d = 11.12 - 13.49$ g/cm^3	−5 to −17 °C	$0.01\ \text{s}^{-1}$	$\sigma_m = A(\theta/\theta_0)^m$	$A = [1.20,\ 1.56]$ MPa, $m = [0.51,\ 0.57]$

(continued)

Table 4 (continued)

Researcher	Specimen	Temperature	Loading/Strain rate	Strength equation	Results
Li et al. [13]	Frozen Lanzhou loess, $\rho = 1.58,\ 1.76$ g/cm^3	-2 to $-15\,°C$	Loose sample: 7×10^{-4} to 1×10^{-6} s^{-1}; dense sample: 6×10^{-4} to 5×10^{-7} s^{-1}	$\sigma_m = A(\theta/\theta_0)^m$	Loose sample: $A = [0.68,\ 1.41]$ MPa, $m = [0.67,\ 0.72]$; dense sample: $A = [0.95,\ 1.53]$ MPa, $m = [0.70,\ 0.75]$
Parameswaran [14]	Saturated frozen Ottawa sand $w = 20\%$	-2 to $-15\,°C$	10^{-7} and 10^{-2} s^{-1}	$\sigma \propto \theta^s$	At higher strain rate $s = 0.44$
Bragg and Andersland [7]	Wedron silica sand with sand volume fraction 64%	$-6\,°C$	1.0×10^{-7} s^{-1} to 1.0×10^{-5} s^{-1}	$\sigma \propto \theta^s$; $\sigma_f \simeq \sigma_{c0}\left(1 + \frac{\theta}{\theta_0}\right)$ [2, 12]	$s = 0.49$ for $\theta < 0\,°C$; linear relationship between σ_f and θ [2, 12]
Yamamoto and Springman [20]	Artificially frozen silty gravel, $w_i = 76 - 84\%$, SG $= 2.71 - 2.72$	-0.3 to -3.0 °C	1.0×10^{-6}, 5.0×10^{-6}, 1.0×10^{-5} s^{-1}	$q = a + bT$, also in [8, 10]	$a = [448.5,\ 1259.7]$ kPa, $b = [-626.5,\ -285.0]$ kPa/°C

1. The magnitude of the exponents in the power function of temperature obtained by all the investigators locates within the range of [0.44, 1.0].
2. The exponent for the coarser-grained frozen soil, e.g. frozen sand, is less than that for the finer-grained frozen soils such as frozen silt and frozen clay.
3. With the increase of strain rate for testing, the exponent in the power function will decrease, while the 'proof strength' (see Eq. 4) will increase.
4. The dry density of frozen soil also has an effect on the exponent, that is, the exponent will increase with the increase of dry density.

Acknowledgements The National Natural Science Foundation of China (grant No. 11702304, 41572268, 41671061), the CAS Pioneer Hundred Talents Program granted to Dr. G. Xu, and the European Commission (grant No. 645665 under Horizon 2020) are acknowledged for their support to this work.

References

1. Andersland, O., Ladanyi, B.: Frozen Ground Engineering. Wiley, URL https://books.google.at/books?id=UyUZ_7o48U0C (2004)
2. Andersland, O.B., Sayles, F.H., Ladanyi, B.: Mechanical Properties of Frozen Ground. McGraw-Hill Book Co. New York, chap. 5, pp. 216–275 (1978)
3. Anderson, D.M., Morgenstern, N.R.: Physics, chemistry, and mechanics of frozen ground: a review. In: North American Contribution Second International Conference on Permafrost, URL https://books.google.at/books?id=EFvHYgEACAAJ (1973)
4. Baker, T. H. W.: Preparation of artificially frozen sand specimens. Tech. Rep. NRCC 15349, National Research Council Canada. Division of Building Research., 15 p (1976)
5. Bourbonnais, J., Ladanyi, B.: The mechanical behavior of a frozen clay down to cryogenic temperatures. In: Kinosita, S., Fukuda, M. (eds.) Proceedings of the 4th International Symposium on Ground Freezing, Rotterdam: A. A. Balkema, Sapporo, Japan, vol. 2, pp. 237–244 (1985)
6. Bourbonnais, J., Ladanyi, B.: The mechanical behavior of frozen sand down to cryogenic temperatures. In: Kinosita, S., Fukuda, M. (eds.) Proceedings of the 4th International Symposium on Ground Freezing, Rotterdam: A. A. Balkema, Sapporo, Japan, vol. 1, pp. 235–244 (1985)
7. Bragg, R.A., Andersland, O.B.: Strain rate, temperature, and sample size effects on compression and tensile properties of frozen sand. Eng. Geol. **18**, 35–46 (1981)
8. Chen, Y.L., Azzam, R., Wang, M., Xu, S., Chang, L.Q.: The uniaxial compressive and tensile tests of frozen saturated clay in shanghai area. Environ. Earth. Sci. **64**(1), 29–36 (2011)
9. Haynes, F.D., Karalius, J.A.: Effect of temperature on the strength of frozen silt, Special report (U.S. Army Cold Regions Research and Engineering Laboratory), vol. 77. U.S. Army Cold Regions Research and Engineering Laboratory, URL http://books.google.at/books?id=Hw8VKmpsLhsC (1977)
10. Hu, X.D., Wang, J.T., Yu, R.Z.: Uniaxial compressive and splitting tensile tests of artificially frozen soils in tunnel construction of hong kong. J. Shanghai. Jiaotong. Univ. (Sci.) **18**(6), 688–692 (2013a)
11. Hu, X.D., Wang, J.T., Yu, X.F.: Laboratory test of uniaxial compressive strength of shanghai frozen soils under freeze-thaw cycle. Mater. Sci. Technol. II, Trans. Tech. Publications, Adv. Mater. Res. **716**, 688–692 (2013b). https://doi.org/10.4028/www.scientific.net/AMR.716.688
12. Ladanyi, B.: An engineering theory of creep of frozen soils. Can. Geotech. J. **9**(1):63–80, http://www.nrcresearchpress.com/doi/pdf/10.1139/t72-005 (1972)

13. Li, H.P., Zhu, Y.L., Pan, W.D.: Uniaxial compressive strength of saturated frozen silt. In: Phillips, M., Springman, S.M., Arenson, L.U. (eds.) International Conference on Permafrost, vol. 8, pp. 679–684 (2003)
14. Parameswaran, V.R.: Deformation behaviour and strength of frozen sand. Can. Geotech. J. **17**(1), 74–88, http://www.nrcresearchpress.com/doi/pdf/10.1139/t80-007 (1980)
15. Sayles, F.H.: Low Temperature Soil Mechanics. Army Cold Regions Research and Engineering Laboratory, Technical note, U.S (1966)
16. Shusherina, E.P., Vyalov, S.S.: Study on the long term strength of frozen soil in uniaxial tests. Moscow State University, Tech. rep. (1963)
17. Tsytovich, N.A.: The Mechanics of Frozen Ground, vol. 10. McGraw-Hill Book Company, New York (1975)
18. Vyalov, S.S.: The strength and creep of frozen soils and calculations for ice-soil retaining structures, Draft translation, vol. 76. U. S. Army Cold Regions Research and Engineering Laboratory, URL http://books.google.at/books?id=0_G-MQAACAAJ (1965)
19. Wijeweera, H., Joshi, R.C.: Compressive strength behavior of fine-grained frozen soils. Can. Geotech. J. **27**(4), 472–483 (1990) http://dx.doi.org/10.1139/t90-062
20. Yamamoto, Y., Springman, S.M.: Axial compression stress path tests on artificial frozen soil samples in a triaxial device at temperatures just below 0 °C. Can. Geotech. J. **51**(10), 1178–1195 (2014). URL http://dx.doi.org/10.1139/cgj-2013-0257, http://dx.doi.org/10.1139/cgj-2013-0257
21. Zhu, Y.L., Carbee, D.L.: Uniaxial compressive strength of frozen silt under constant deformation rates. Cold. Reg. Sci. Technol. **9**(1), 3–15 (1984). https://doi.org/10.1016/0165-232X(84)90043-0, URL http://www.sciencedirect.com/science/article/pii/0165232X84900430

The manufacturer's authorised representative in the EU is Springer Nature Customer Service Centre GmbH, Europaplatz 3, 69115 Heidelberg, Germany. If you have any concerns regarding our products, please contact ProductSafety@springernature.com

Printed and bound by CPI Group (UK) Ltd, Croydon, CR0 4YY
05/10/2026
02228141-0003